B Tindale

MULTIPLE-CHOICE QUESTIONS
IN BIOLOGY AND HUMAN BIOLOGY

MULTIPLE-CHOICE QUESTIONS IN BIOLOGY AND HUMAN BIOLOGY

O. F. G. KILGOUR
B.Sc., M.I.Biol., A.I.F.S.T.

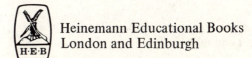
Heinemann Educational Books
London and Edinburgh

Heinemann Educational Books Ltd

LONDON EDINBURGH MELBOURNE AUCKLAND TORONTO
HONG KONG SINGAPORE KUALA LUMPUR
IBADAN NAIROBI JOHANNESBURG
LUSAKA NEW DELHI

Owen Jones
. . . his memory

Published by Heinemann Educational Books Ltd.,
48 Charles Street, London W1X 8AH

Printed in Great Britain by The Whitefriars Press Ltd.,
London and Tonbridge

Preface

As the worktime in schools comes under increasing pressure from syllabuses, far less time remains available for a thorough testing of the pupils' knowledge. Multiple-choice type questions have come to be regarded as the most useful and versatile form of assessing a pupil's progress, as they can be used both for continuous assessment, and for revision and examination purposes.

The questions in this book cover the Biology and Human Biology syllabuses of most examinations at GCE 'O' and CSE levels. Sections 1–9, and 17–18 also provide useful tests for nursing students who are preparing for the GNC Part 1 examinations.

Acknowledgements

I am indebted to Brian Bracegirdle, Patricia Miles, and W. H. Freeman for the use of photographic material, and also wish to express my appreciation for the help and encouragement received from my publishers.

O.F.G.K.

Contents

1

Living Things, Cells, and Tissues

Questions 1 and 2 refer to Figure 1.1:

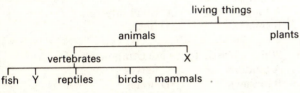

Figure 1.1

Select your answer from the following:
A amphibia
B cabbages
C invertebrates
D toadstools.

1   Which group is placed at Y?

2   Which group is placed at X?

Questions 3 and 4 refer to Figure 1.2:

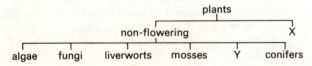

Figure 1.2

Select your answer from the following:
A ferns
B flowering
C seaweeds
D pine trees.

3   Which group is placed at X?

4   Which group is placed at Y?

5   Which of the following is a cold-blooded vertebrate?
A penguin
B turtle
C seal
D whale.

6   Which of the following is a warm-blooded vertebrate?
A shark
B eel
C porpoise
D tortoise.

7   A body surface *completely* covered in scales is a feature of:
A lizards
B cats
C pigeons
D newts.

8   A body surface which is only *partly* covered in scales is a feature of:
A herrings
B hens
C frogs
D snakes.

9   Refer to Figure 1.3:

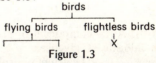

Figure 1.3

Which of the following would be placed at position X?
A ostrich
B swallow
C eagle
D sparrow.

10   Which two of the following are components of animal cells?
(i) cellulose walls, (ii) small vacuoles, (iii) sodium chloride, (iv) chloroplasts.
A (ii), (iii)
B (i), (ii)
C (iii), (iv)
D (i), (iv)

11   Which two of the following are classed as plants?
(i) moulds, (ii) amoeba, (iii) yeast, (iv) coral.
A (ii), (iii)   C (i), (ii)
B (i), (iii)   D (ii), (iv)

12   Which three of the following are classed as animals?
(i) toadstools, (ii) sponges, (iii) snails, (iv) cockles.
A (i), (ii), (iii)   C (i), (iii), (iv)
B (ii), (iii), (iv)   D (i), (ii), (iv)

13   Which two of the following are invertebrate animals?
(i) tadpoles, (ii) bees, (iii) crabs, (iv) sea horses.
A (i), (ii)   C (ii), (iii)
B (iii), (iv)   D (i), (iv)

14   Select examples of three reptiles from the following:
(i) slow-worm, (ii) adder, (iii) pheasant, (iv) lizard.
A (i), (ii), (iii)   C (ii), (iii), (iv)
B (i), (ii), (iv)   D (i), (iii), (iv)

15   Select two conifer trees from the following:
(i) larch, (ii) holly, (iii) spruce, (iv) apple.
A (i), (iii)   C (ii), (iv)
B (ii), (iii)   D (iii), (iv)

16   Which two of the following are insects?
(i) shrimps, (ii) lobster, (iii) ants, (iv) wood-worm.
A (i), (ii)   C (iii), (iv)
B (ii), (iii)   D (i), (iv)

17   The main structural component of plant cells is:
A sodium chloride   C glycogen
B cellulose   D calcium phosphate.

18   The part of the animal cell concerned with energy
producing reactions is called the:
A nucleolus   C chloroplast
B centrosome   D mitochondria.

19   A tissue which covers and lines mammalian body
surfaces is called:
A epithelium   C bone
B skin   D muscle.

20   Which of the following is a *system* of the mammalian
body?
A heart   C stomach
B liver   D skeleton.

21   Ciliated columnar epithelium of a mammal lines the:
A mouth   C air passages
B small intestine   D body cavity.

22   Pavement or squamous tissue is found as a covering
or lining of the:
A lung air sacs   C small intestine
B skin   D nose.

23   White inelastic fibres are components of:
A tendons   C nerves
B kidneys   D bones.

24   The iris of the eye is controlled by a muscle which is:
A voluntary   C involuntary
B striated   D cardiac.

25   The type of muscle forming the tongue and first part
of the throat, which are used in the process of swallowing,
are called:
A voluntary   C involuntary
B smooth   D unstriped.

26   A lack or breakdown in the supply of one of the
following, for a period of more than 10 minutes, will
cause permanent damage to nerve cells:
A food   C oxygen
B water   D salt.

27   Which of the following tissues can change their shape,
or contract?
A cartilage   C ligaments
B cardiac muscle   D red blood cells.

28   What type of muscle causes movement of the stomach
wall?
A smooth   C voluntary
B striped   D cardiac.

29 Ligaments function to connect together:
    A bones   C muscles to bones
    B muscles   D intestines.

30 Tendons function to connect together:
    A bones   C muscles to bones
    B muscles   D intestines.

31 Tendons consist mainly of large amounts of tissue called:
    A white inelastic   C yellow elastic
       fibres   fibres
    B cartilage   D areolar.

32 The axon fibre of a neurone has a covering or sheath of:
    A white fibres   C amylose
    B myelin   D perimysium.

Questions 33–37 refer to Figure 1.4 which shows isolated single cells that line or cover the mammalian body.

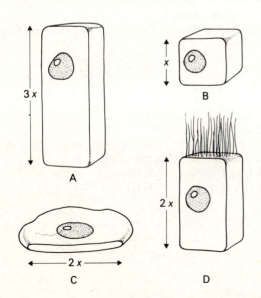

Figure 1.4

33 Which cell has the greatest surface area?

34 Which cell is most suited for the purpose of gas and liquid diffusion?

35 Which cell would drive mucus along the air passages towards the mouth?

36 Which of the non-ciliated cells would produce the greatest volume of secreted fluid?

37 The four cells in Figure 1.4 are examples of what type of tissue?
    A connective   C epithelial
    B muscle   D nerve.

38 The type of epithelium lining blood vessels is called:
    A columnar   C pavement
    B ciliated   D cubical.

Questions 39–41 refer to Figure 1.5.

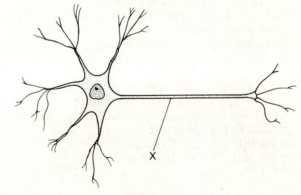

Figure 1.5

39 Figure 1.5 illustrates the structure of:
    A amoeba   C segmented worm
    B neurone   D root hair.

40   The part labelled X in Figure 1.5 is called the:
A septum   C cell vacuole
B pseudopodium   D axon.

41   The function of the part labelled X is:
A movement   C conductivity
B food passage   D reproduction.

42   Which two of the following are examples of
individual or single tissues?
(i) lungs, (ii) blood, (iii) fat, (iv) heart muscle.
A (i), (ii)   C (iii), (iv)
B (ii), (iii)   D (i), (iv)

43   Which two of the following function as connective
tissue?
(i) biceps muscle, (ii) axon fibres, (iii) adipose tissue,
(iv) tendons.
A (i), (ii)   C (ii), (iii)
B (iii), (iv)   D (i), (iv)

44   Which three of the following in the mammalian body
contain smooth involuntary muscle?
(i) intestine, (ii) triceps muscle, (iii) blood vessels,
(iv) urinary bladder.
A (i), (iii), (iv)   C (ii), (iii), (iv)
B (i), (ii), (iii)   D (i), (ii), (iv)

45   Which two of the following mammalian body parts is
composed of striped voluntary muscle?
(i) masseter muscle, (ii) bronchial muscle, (iii) ciliary
muscle, (iv) biceps muscle.
A (i), (ii)   C (iii), (iv)
B (ii), (iii)   D (i), (iv)

46   The three functions of adipose tissue are:
(i) food storage, (ii) shock absorbtion, (iii) muscle
connection, (iv) heat insulation.
A (i), (ii), (iv)   C (i), (ii), (iii)
B (ii), (iii), (iv)   D (i), (iii), (iv)

47   Pavement or squamous cells line two of the following:
(i) lung air sacs, (ii) arteries and veins, (iii) stomach,
(iv) bronchial tubes.
A (iii), (iv)   C (i), (iv)
B (ii), (iii)   D (i), (ii)

48   Which two of the following are component parts of a
neurone?
(i) cortex, (ii) dendrites, (iii) axon, (iv) radicle.
A (i), (ii)   C (iii), (iv)
B (ii), (iii)   D (i), (iv)

49   Which three of the following are organs of the
mammalian body?
(i) salivary glands, (ii) masseter muscles, (iii) finger nails,
(iv) ears.
A (i), (ii), (iv)   C (i), (ii), (iii)
B (ii), (iii), (iv)   D (i), (iii), (iv)

50   Which two of the following tissues are present in the
structure of a finger?
(i) hyaline cartilage, (ii) striated muscle, (iii) ciliated
epithelium, (iv) choroid tissue.
A (iii), (iv)   C (i), (ii)
B (ii), (iii)   D (i), (iv)

2

Skeleton, Joints, and Movement

1   A jellyfish obtains support for its body from:
    A the endoskeleton   C cell thickening
    B cell swelling   D the exoskeleton.

2   Which one of the following has an exoskeleton?
    A snail   C hydra
    B earthworm   D tapeworm.

Questions **3** and **4** refer to Figure 2.1.

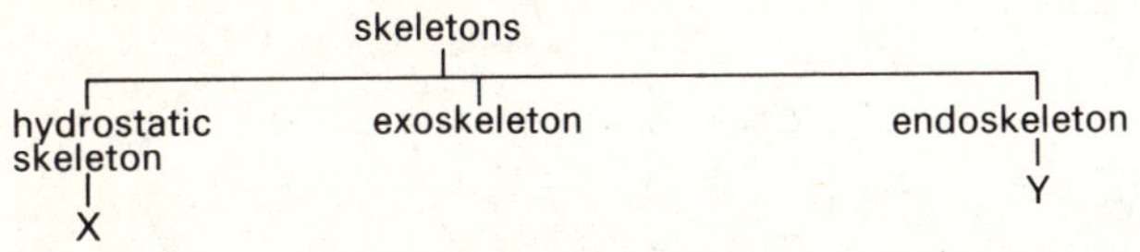

Figure 2.1

Answer the questions using the following possible answers:
    A earthworm   C rabbit
    B cockle   D beetle.

3   Which animal would have a body support of the type X?

4   Which animal would have a body support of the type Y?

Questions **5** and **6** refer to Table 2.1.

5   An example of X is:
    A phalanges   C mandible
    B femur   D carpals

6   An example of Y is:
    A scapula   C tibia
    B patella   D clavicle.

TABLE 2.1

Types of bones	Examples
long	X
short	Tarsals
flat	Y
irregular	Vertebrae

7   The first vertebra of the mammalian vertebral column immediately next to and supporting the skull is called the:
    A axis   C sacrum
    B atlas   D thoracic.

8   Which one of the following vertebrae are connected through movable joints with the ribs?
    A sacral   C cervical
    B lumbar   D thoracic.

9   Which one of the following parts of the rib cage is anterior to the backbone?
    A floating rib   C scapula
    B sternum   D true rib.

10 The pelvic girdle is firmly attached to a part of the vertebral column called the:
    A atlas   C sacrum
    B coccyx   D axis.

11 The pectoral girdle is attached by muscles to a part of the trunk called the:
A thorax   C pelvis
B sacrum   D sternum.

Questions **12–15** refer to the diagram of the arm in Figure 2.2.

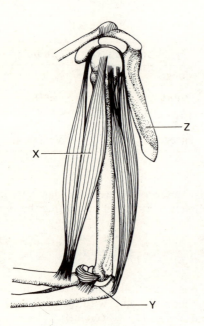

Figure 2.2

12 What is the name of the muscle X?
A masseter   C biceps
B triceps   D extensor.

13 Name the type of joint at Y.
A hinge   C ball and socket
B pivot   D gliding.

14 What is the name of the portion of the girdle labelled Z?
A clavicle   C humerus
B sternum   D scapula.

15 To what part of the skeleton is the *insertion* portion of muscle X attached?
A humerus   C radius
B ulna   D scapula.

16 Which one of the following will cause a straightened arm to bend due to muscle contraction?
A triceps   C biceps
B extensors   D quadriceps.

17 The bones of the wrist are called the:
A tarsals   C metatarsals
B carpals   D metacarpals.

Questions **18-20** refer to the lower limb skeleton shown in Figure 2.3.

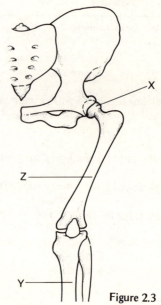

Figure 2.3

18 Name the type of joint at X.
A synovial   C immovable
B fibrous   D cartilaginous.

19 What is the name of the long bone Y?
A patella   C fibula
B femur   D tibia.

20 The part of the bone labelled Z is known as the:
A shaft   C neck
B head   D condyle.

21 Which one of the following body joints have all these functions or features:
(i) great mobility, (ii) easily dislocated, (iii) unable to bear great weight.
A hip   C ankle
B shoulder   D knee.

22 Which one of the following is an example of immovable bone joints?
A wrist   C cranium bones
B elbow   D ankle.

23 The centrum, transverse process, neural arch, and neural canal are component parts of the:
A pelvic girdle   C thoracic vertebrae
B scapula   D humerus.

24 Which one of the following body joints is lubricated by a synovial fluid?
A skull sutures   C between finger phalanges
B between rib and   D tooth sockets.
   sternum

Questions **25** and **26** refer to Table 2.2.

TABLE 2.2

Type of joint	Example
hinge	between humerus and ulna
pivot	X
ball and socket	between femur and acetabulum
gliding	Y

25 Which of the following is an example of X?
A atlas/axis   C radius/humerus
B scapula/humerus   D cranium bones.

26 Which of the following is Y?
A between toe   C between finger
   phalanges   phalanges
B head of radius   D between carpal
   ulna   bones.

27 Which of the following is *not* a function of the mammalian skeleton?
A blood cell   C calcium storage
   formation
B protection of   D glycogen storage.
   soft organs

28 Which two of the following animals have an exoskeleton?
(i) snake, (ii) crab, (iii) fish, (iv) cockroach.
A (i), (iii)   C (i), (ii)
B (iii), (iv)   D (ii), (iv)

29 Which two of the following animals have an endoskeleton?
(i) jellyfish, (ii) lobster, (iii) frog, (iv) fish.
A (iii), (iv)   C (i), (iii)
B (i), (ii)   D (ii), (iv)

30 Which two of the following can undergo the process of ecdysis?
(i) rabbit, (ii) shrimp, (iii) butterfly, (iv) rat.
A (ii), (iii)   C (i), (iii)
B (i), (ii)   D (ii), (iv)

31 The two parts concerned with growth in length and girth of a long bone are the:
(i) articulate condyles, (ii) epiphysis, (iii) periosteum, (iv) shaft.
A (i), (ii)   C (i), (iv)
B (ii), (iii)   D (iii), (iv)

32 Starting with the vertebrae nearest to the skull arrange the following vertebrae in their correct order as they appear in the mammalian vertebral column.
(i) sacral, (ii) thoracic, (iii) lumbar, (iv) cervical.
A (ii), (iv), (iii), (i)   C (iii), (iv), (ii), (i)
B (i), (ii), (iii), (iv)   D (iv), (ii), (iii) (i)

33 Which two of the following are bones of the forearm?
(i) humerus, (ii) clavicle, (iii) radius, (iv) ulna.
A (ii), (iii)   C (i), (ii)
B (iii), (iv)   D (i), (iv)

34 Which two of the following are bones of the lower leg?
(i) tibia, (ii) femur, (iii) ilia, (iv) fibula.
A (i), (iv)   C (ii), (iv)
B (i), (iii)   D (ii), (iii)

35 Which three of the following are component parts of a femur?
(i) head, (ii) centrum, (iii) condyles, (iv) neck.
A (i), (ii), (iii)   C (i), (ii), (iv)
B (ii), (iii), (iv) D (i), (iii), (iv)

36 Which two of the diagrams in Figure 2.4 are examples of third order levers in the human body?
A 1, 2   C 3, 4
B 2, 4   D 1, 3

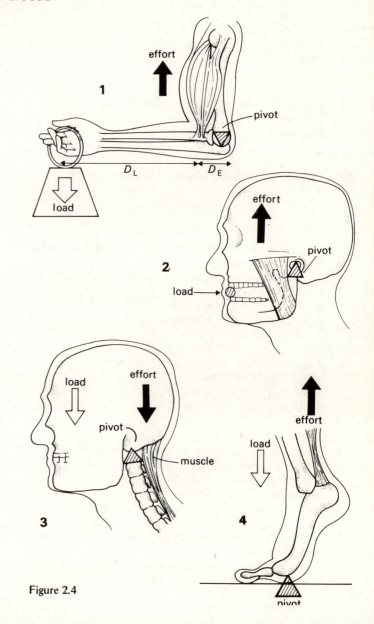

Figure 2.4

37 Which bones of the human skull are movable?
(i) hard palate, (ii) lower jaw, (iii) ear ossicles, (iv) upper jaw.

A (i), (ii)   C (ii), (iii)
B (iii), (iv)   D (i), (iv)

38 Which three of the following vertebrates have pentadactyl limbs?
(i) bat, (ii) pigeon, (iii) whale, (iv) herring.

A (i), (ii), (iv)   C (ii), (iii), (iv)
B (i), (ii), (iii)   D (i), (iii), (iv)

39 Which of the following muscles are antagonistic pairs?
(i) triceps, (ii) biceps, (iii) quadriceps, (iv) masseters.

A (i), (iv)   C (iii), (iv)
B (ii), (iii)   D (i), (ii)

40 Which three of the following are diseases of bones and joints?
(i) rheumatism, (ii) bronchitis, (iii) rickets, (iv) arthritis.

A (i), (ii), (iii)   C (ii), (iii), (iv)
B (i), (iii), (iv)   D (i), (ii), (iv)

41 The large muscles of the buttocks are called:

A gluteal muscles   C masseter muscles
B pectoralis muscles   D deltoid muscles.

42 The tendon of Achilles is joined to the calf muscle called the:

A sartorius   C gastrocnemius
B quadriceps   D biceps.

43 Which three of the following are muscles of the upper limb?
(i) quadriceps femoris, (ii) triceps, (iii) deltoid, (iv) brachialis.

A (ii), (iii), (iv)   C (i), (ii), (iv)
B (i), (ii), (iii)   D (i), (iii), (iv)

44 Which three of the following are muscles of the trunk?
(i) gastrocnemius, (ii) latissimus dorsi, (iii) erector spinae, (iv) pectoralis major.

A (i), (ii), (iii)   C (i), (ii), (iv)
B (ii), (iii), (iv)   D (i), (iii), (iv)

45 Which three of the following are muscles of the head and neck?
(i) masseter, (ii) orbicularis oris, (iii) sterno mastoid, (iv) latissimus dorsi

A (ii), (iii), (iv)   C (i), (ii), (iii)
B (i), (ii), (iv)   D (i), (iii), (iv)

3

Food and the Digestive System

1 Which one of the following groups of elements make-up sugars and lipids or fats?
A. Na, Cl
B. C, H, O, N, S
C. Ca, K, Mg, Fe
D. C, H, O.

2 Select one of the following food groups which is composed of the chemical elements carbon, hydrogen, oxygen, sulphur, nitrogen, and phosphorus.
A protein
B lipid or fat
C carbohydrates
D water.

3 Which one of the following food nutrients contains the greatest amount of carbon, compared to other component elements.
A protein
B lipid or fat
C carbohydrate
D vitamins.

4 Which nutrient is provided in excessive amounts by toffee, lemonade, honeycomb, and treacle?
A carbohydrate
B lipid or fat
C proteins
D vitamins.

5 Select a polysaccharide from the following:
A glucose
B sucrose
C cellulose
D lactose.

6 Which one of the following is split up by the chemical action of water, or hydrolysis, during digestion?
A glucose
B fructose
C invert sugar
D sucrose.

7 Select the foodstuff which contains the largest amount of cellulose.
A cane sugar
B cabbage
C lean beef steak
D cheese.

8 Which one of the following is called blood sugar?
A starch
B glycogen
C glucose
D lactose.

9 One of the following is insoluble in water.
A glucose
B fructose
C maltose
D cellulose.

10 Which one of the following is an edible lipid oil?
A diesel oil
B peanut oil
C paraffin oil
D clove oil.

11 Select the foodstuff from the following with the highest lipid content:
A bread
B milk
C butter
D apple.

12 Lipid oils and fats are composed of substances called:
A fatty alcohols and glycerine
B fatty acids and glycogen
C fatty acids and glycerine
D fatty alcohols and glucose.

13 Which one of the following foodstuffs contains the largest amount of protein?
A white fish
B white bread
C milk
D potatoes.

14 Proteins are built up of simpler units called:
A fatty acids   C monosaccharides
B glycerides   D amino acids.

15 Which one of the following foods contains protein of the highest biological value?
A gelatine   C peas
B eggs   D wholemeal bread.

16 Which one of the following food components contains the chemical element nitrogen?
A glucose   C starch
B olive oil   D egg albumen.

17 Which one of the following nutrients, needed for the purpose of tissue growth and repair, cannot be made by the human body?
A vitamin D   C glycogen
B essential amino acids D essential oils.

18 Select the foodstuff containing the mineral element essential for making the red pigment of blood cells.
A milk   C cheese
B cane sugar   D liver.

19 Which one of the following foods contains good sources of the mineral elements necessary for forming bones and teeth?
A butter   C table jelly
B cheese   D pears.

20 One of the following foods provides the mineral element needed for the healthy functioning of the thyroid gland.
A oranges   C sea fish
B rice   D kidneys.

21 In certain parts of the country, drinking water is treated with one of the following to prevent tooth decay:
A chlorides   C sulphates
B fluorides   D bicarbonates.

22 Cows milk is unsuitable as food for human babies because it lacks one of the following nutrients:
A iron   C phosphorus
B calcium   D vitamin C.

23 Select one of the following body secretions which contains iodine:
A saliva   C bile juice
B thyroxine   D insulin.

24 Rickets is a disease in children preventable by eating adequate amounts of one of the following foodstuffs daily:
A fruit   C green vegetables
B margarine   D rice.

25 Scurvy is a disease due to a deficiency of one of the following nutrients from the daily diet:
A vitamin C   C iron
B vitamin A   D fluoride.

26 Which one of the following foodstuffs would produce a dark blue to black colour when treated with a few drops of iodine solution?
A icing sugar   C table salt
B white bread   D dried milk powder.

27 One of the following substances will produce a reddish brown colour when boiled gently with Benedict's or Fehling's solution.
A gelatine   C corn oil
B vinegar   D honey.

28 Which of the following groups of vitamins are found in foods of a high lipid or fat content?
A. vitamins B and C   C. vitamins A, C, and E
B. vitamins of the B   D. vitamins A, D, and E.
   group

29   Which one of the following vitamins dissolve readily in cooking water when food is cooked by boiling or stewing?
A vitamin A
B vitamin B
C vitamin D
D vitamin E.

30   Taking equal weights of the following *pure* nutrients, which one will produce the greatest heat energy value on burning or combustion?
A carbohydrate
B protein
C lipid
D alcohol.

31   Which one of the following exercises would use up the greatest amount of energy, if performed continually for 15 minutes?
A walking upstairs
B walking on a level road
C walking up a gentle slope
D walking downstairs.

32   Which of the following groups of food could be excluded from a balanced daily diet?
A sugar, jam, confectionery
B fresh fruit
C fish, meat, eggs, and dairy products
D vegetables.

33   Which one of the following provides roughage for healthy intestinal action?
A glucose
B lard
C cellulose
D vitamin C

34   Select a food which can provide roughage in a diet.
A cabbage
B roast beef
C steamed white fish
D poached eggs.

35   Which three of the following nutrients can provide the body with energy?
(i) carbohydrates, (ii) proteins, (iii) lipids, (iv) minerals.
A (i), (ii), (iii)
B (ii), (iii), (iv)
C (i), (ii), (iv)
D (i), (iii), (iv)

36   Which of the following nutrients are needed for body building and tissue repair?
(i) carbohydrates, (ii) vitamins, (iii) proteins, (iv) minerals.
A (i), (ii)
B (ii), (iii)
C (i), (iii)
D (iii), (iv)

37   Which of the following nutrients protect the body from disease and regulate metabolic activities?
(i) carbohydrates, (ii) minerals, (iii) vitamins, (iv) lipids.
A (i), (ii)
B (ii), (iii)
C (iii), (iv)
D (ii), (iv)

Questions 38 and 39 refer to Figure 3.1.

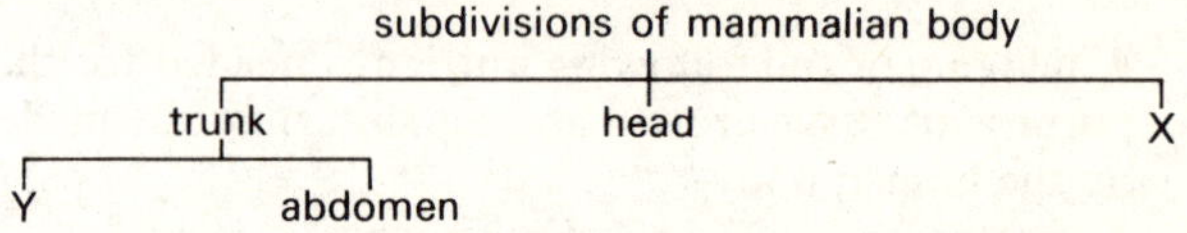

Figure 3.1

Answer the questions using the following possible answers:
A neck
B thorax
C face
D coelom.

38   Which name should be placed at X?

39   Which name should be placed at Y?

40   Which one of the following organs are located in the abdominal cavity?
A liver
B heart
C lungs
D thymus gland.

41   Which one of the following organs are located in the pelvic cavity of the abdomen?
A stomach
B bladder
C pancreas
D duodenum.

42   Which one of the following animals has a thorax?
A herring
B honey bee
C earthworm
D snail.

43 Which one of the following animals has a diaphragm?
A dogfish   C frog
B earthworm   D rabbit.

44 Which two of the following are components of the thoracic cavity of a mammal?
(i) aorta, (ii) spleen, (iii) oesophagus, (iv) pancreas.
A (i), (iii)   C (i), (ii)
B (ii), (iv)   D (iii), (iv)

45 Which type of human tooth cuts the first bite out of an apple?
A premolar   C incisor
B canine   D molar.

46 The crown of a tooth is capped with a very hard substance called:
A enamel   C dentine
B cement   D plaque.

47 Which one of the following animals have canine teeth?
A hare   C cat
B mouse   D rabbit.

Questions **48** and **49** refer to Figure 3.2.

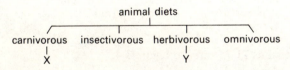

Figure 3.2

Select your answers from the following:
A hedgehog   C lion
B rabbit   D mouse.

48 Which animal has the diet grouped at X?

49 Which animal has the diet grouped at Y?

50 The changing of starch into maltose can take place in the:
(i) mouth, (ii) large intestine, (iii) stomach, (iv) small intestine.
A (i), (ii)   C (i), (iv)
B (ii), (iii)   D (iii), (iv)

51 Which two of the following have sphincter muscles?
(i) duodenum, (ii) tongue, (iii) liver, (iv) stomach.
A (i), (iv)   C (ii), (iii)
B (i), (ii)   D (ii), (iv)

52 The food nutrient acted upon by stomach enzymes is:
A lipid   C minerals
B carbohydrate   D protein.

53 Which one of the following components of gastric juice serves as a food disinfectant?
A mucus   C hydrochloric acid
B pepsin   D rennin.

54 Which one of the following digestive juices is without an enzyme content?
A gastric juice   C bile juice
B pancreatic juice   D saliva.

55 Lipids are emulsified into easily digested emulsions by the action of:
A mucin   C rennin
B bile salts   D pepsin.

56 Which one of the following can to some extent digest all of the nutrients, carbohydrates, lipids, and proteins?
A gastric juice   C intestinal juice
B bile juice   D saliva.

57 Which one of the following secretions of the pancreas passes directly into the blood?
A trypsinogen   C insulin
B amylase   D lipase.

58   Which one of the following digestive juices is unable to digest carbohydrates?
A saliva
B gastric juice
C pancreatic juice
D intestinal juice.

59   In which one of the following is a villus and lacteal vessel found?
A small intestine
B large intestine
C stomach
D liver.

60   Which one of the following enter the lacteal vessel of villi?
A glycogen
B glycerine
C glucose
D sucrose.

61   The villi are finger shaped in order to:
A move food along the intestine
B penetrate the chyme
C provide a large absorption surface
D sift out undigested foods.

62   Which two of the following show peristaltic movements?
(i) small intestine, (ii) ureters, (iii) tongue, (iv) eyelids.
A (i), (ii)
B (i), (iii)
C (iii), (iv)
D (i), (iv)

63   Which one of the following is absorbed into the lacteal vessel of a villus?
A amino acids
B fatty acids
C hydrochloric acid
D ascorbic acid.

64   Which one of the following is unable to secrete digestive juices or enzymes?
A stomach
B duodenum
C ileum
D colon.

65   In which part of the alimentary canal are vitamins B and K made by the activity of certain bacteria?
A stomach
B duodenum
C ileum
D colon.

66   The purpose of the colon is to absorb one of the following from the intestine:
A mucin
B water
C cellulose
D bile pigments.

Questions **67** and **68** refer to Figure 3.3.

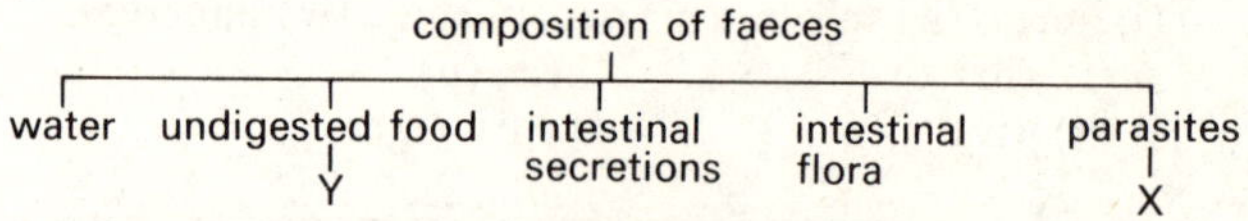

composition of faeces

water   undigested food   intestinal secretions   intestinal flora   parasites
Y
X

Figure 3.3

67   Which of the following would be placed at Y?
A mucus
B bacteria
C cellulose fibres
D enzymes.

68   Which of the following would be placed at X?
A louse
B threadworm
C fungi
D ringworm.

69   Which of the following body fluids from a healthy person would produce a reddish brown colour on boiling with Benedicts or Fehlings solution?
A saliva
B sweat
C urine
D plasma.

70   Which of the following is the largest body gland found close to the dome of the diaphragm?
A pancreas
B thymus
C spleen
D liver.

Questions **71-75** refer to Figure 3.4.

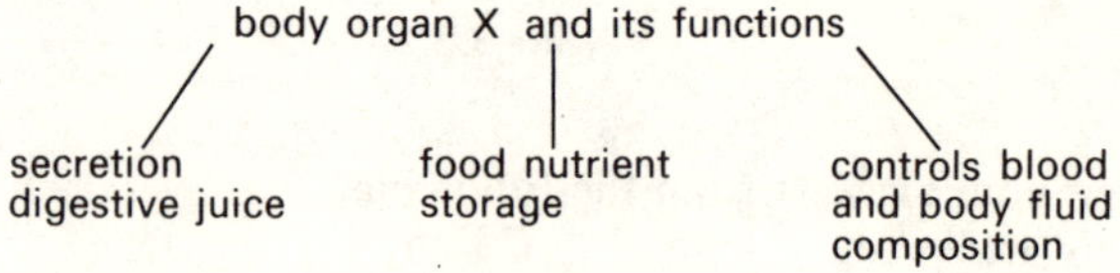

body organ X and its functions

secretion digestive juice   food nutrient storage   controls blood and body fluid composition

Figure 3.4

71   Which one of the following would be body organ X?
A pancreas
B liver
C spleen
D duodenum.

72   Name the carbohydrate stored in this organ X:
      A glycogen   C glucose
      B cellulose   D amylose.

73   What mineral element is stored in the organ X?
      A calcium   C iron
      B fluoride   D iodine.

74   Which vitamin essential for making new red blood cells
      is found in this organ?
      A. vitamin A   C. vitamin B_1
      B. vitamin D_2 D. vitamin B_{12}.

75   Deamination occurs in organ X when amino acids are
      changed to glucose and a harmless excretory product
      called:
      A ammonia   C uric acid
      B urea   D sodium chloride.

76   Which organ makes a hormone secretion that controls
      the use of glucose in the body?
      A stomach   C pancreas
      B liver   D spleen.

4

Circulatory System

1 One of the following does not possess a circulatory system.
   A frog   C earthworm
   B amoeba   D snail.

2 The liquid bathing and linking the life activities of body cells together is called:
   A aqueous humour   C mucus
   B tissue fluid   D serum.

Questions 3–12 refer to Figure 4.1.

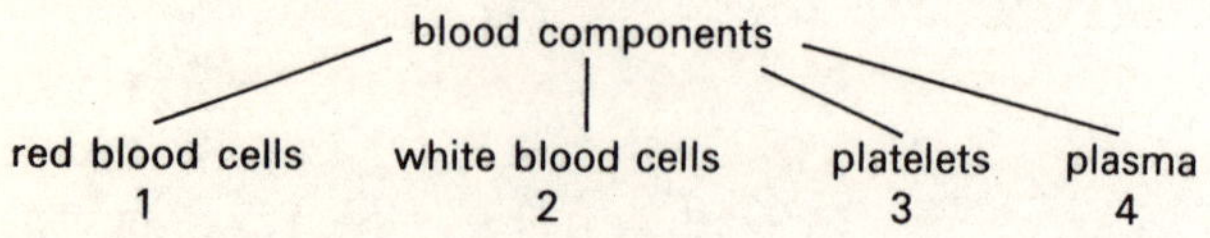

blood components

red blood cells   white blood cells   platelets   plasma
     1   2   3   4

Figure 4.1

3 Haemoglobin is a component of:
   A 1   C 2
   B 3   D 4.

4 A network of fibrin is formed in a blood clot by:
   A 1   C 2
   B 3   D 4.

5 Which labelled component fights infection in a septic wound?
   A 1   C 2
   B 3   D 4.

6 The blood component which is 92 per cent water is:
   A 1   C 2
   B 3   D 4.

7 Which is the most abundant cellular component of blood?
   A 1   C 2
   B 3   D 4.

8 The largest blood cell that also has a nucleus is:
   A 1   C 2
   B 3   D 4.

9 Amino acids, glucose, and salts are found in:
   A 1   C 2
   B 3   D 4.

10 Iron is needed for the healthy functioning of:
   A 1   C 2
   B 3   D 4.

11 The blood component involved, or affected by, the disease called anaemia is:
   A 1   C 2
   B 3   D 4.

12 Which labelled component has the power of amoeboid movement?
   A 1   C 2
   B 3   D 4.

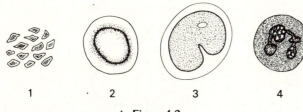

Figure 4.2

13 Select the diagram from Figure 4.2 which resembles the structure of a red blood cell.
A 1   C 2
B 3   D 4.

14 Viewed through a microscope, what will be the effect on the red blood cells when a drop of distilled water is added to a drop of blood?
A shrink   C dissolve
B swell   D not change.

15 Seen through a microscope, what will be the effect on the red blood cells when a drop of strong solution of salt (sodium chloride) is added to a drop of blood?
A shrink   C dissolve
B swell   D not change.

16 Which three of the following take part in the process of blood clotting?
(i) amino acids, (ii) calcium, (iii) platelets, (iv) prothrombin.
A (i), (ii), (iii)   C (i), (iii), (iv)
B (ii), (iii), (iv)   D (i), (ii), (iv)

Questions 17 and 18 refer to Figure 4.3.

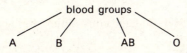

Figure 4.3

17 A person called a universal recipient has blood group type:
A. A   C. O
B. B   D. AB.

18 A person called a universal donor has blood group type:
A. A   C. O
B. B   D. AB.

19 Red blood cells are made in the:
A liver   C bone marrow
B spleen   D lymph glands.

20 Carbon dioxide is carried in the blood by:
(i) plasma, (ii) white blood cells, (iii) platelets, (iv) red blood cells.
A (i), (ii)   C (i), (iv)
B (ii), (iv)   D (ii), (iii).

21 Human blood provides immunity against certain diseases by producing:
A antibiotics   C antibodies
B vaccines   D anticoagulants.

22 Immunity against certain diseases can be obtained by injections of dead or treated bacteria or viruses called:
A sera   C antibodies
B vaccines   D antibiotics.

23 Which one of the following animals has a heart contained within a distinct thorax?
A frog   C bird
B fish   D rat.

24 Surrounding the heart of a mammal is a double-layered, fluid-filled, membranous bag called the:
A pleura   C pericardium
B pericarp   D peritoneum.

25 Which one of the following has a heart?
A amoeba   C spirogyra
B water flea   D seaweed.

26   Which one of the following layers of the heart wall is in actual contact with the blood flowing through the heart chambers?
A myocardium   C endocardium
B epicardium   D pericardium.

27   The artery which supplies the heart with blood is called the:
A carotid   C aorta
B coronary   D pulmonary.

28   Auricles are parts of one of the following:
A heart   C lungs
B intestine   D liver.

Questions 29–31 refer to Table 4.1.

TABLE 4.1

Heart chamber	Great blood vessel adjacent to chamber	Blood oxygen and food composition
Right atrium	X	Y
Left ventricle	Z	Oxygenated

29   The largest blood vessel adjacent to the right atrium or X is:
A aorta   C vena cava
B pulmonary artery   D portal vein.

30   The oxygen and food content in the blood of Y is:
A oxygenated minus food   C deoxygenated plus food
B deoxygenated minus food   D oxygenated plus food.

31   The blood vessel Z adjacent to the left ventricle is:
A pulmonary vein   C pulmonary artery
B aorta   D portal vein.

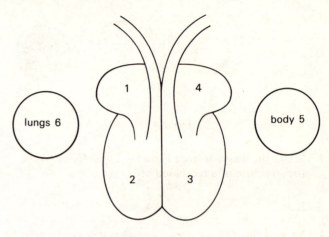

Figure 4.4

32   Select the correct sequence of blood flowing through the heart commencing with the left ventricle shown in Figure 4.4.

A 4, 3, 2, 6, 1, 5   C 3, 5, 1, 2, 6, 4
B 4, 3, 6, 1, 2, 5   D 3, 5, 6, 4, 2, 1.

33   Red blood cells live about 120 days and are then destroyed in the:
A stomach   C spleen
B kidneys   D bone marrow.

34   Which one of the following heart chambers or vessels contains deoxygenated blood?
A right atrium   C pulmonary vein
B left atrium   D left ventricle.

35   Blood vessels undergo constriction during one of the following:
A blushing   C hot baths
B ice-pack treatment D heat-ray treatment.

36 Which three of the following parts of the heart contain, or come into close contact, with oxygenated blood?
(i) mitral bicuspid valve, (ii) tricuspid valve, (iii) pulmonary artery, (iv) aorta, (v) right atrium, (vi) left ventricle.
A (i), (iii), (iv)   C (i), (iv), (vi)
B (ii), (iii), (v)   D (ii), (iv), (v)

37 Which one of the following are valves found between the right atrium and the right ventricle?
A bicuspid   C tricuspid
B semi-lunar   D mitral.

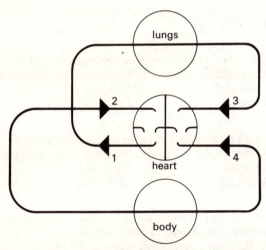

Figure 4.5

38 Which arrow wrongly indicates the flow of mammalian blood in Figure 4.5.

A 4   C 2
B 1   D 3.

39 During the process of *one heart beat* the following events take place:
(i) atrial systole, (ii) atrial diastole, (iii) ventricular systole, (iv) ventricular diastole.

In which order of events does *contraction* of the heart occur as opposed to relaxation?
A (i), (ii)   C (i), (iv)
B (i), (iii)   D (ii), (iii)

40 What is the average pulse or heart beat per minute of a new-born baby?
A 60   C 72
B 140   D 85.

41 Which of the following *slows down* the heart beat rate or pulse?
A exercise   C shock
B excitement   D pain.

42 Which of the following will have the effect of lowering the blood pressure?
A sleep   C exercise
B excitement   D deep-sea diving.

43 Select the correct order of blood flow through the following *from* the heart:
(i) veins, (ii) arterioles, (iii) venules, (iv) arteries, (v) capillaries.
A (i), (iii), (ii), (v), (iv)   C (iv), (ii), (v), (iii), (i)
B (i), (iii), (ii), (iv), (v)   D (iii), (i), (ii), (v), (iv)

44 Select the correct arrangement of the following in order of decreasing muscle wall thickness.
(i) arteries, (ii) capillaries, (iii) heart, (iv) veins.
A (ii), (iv), (iii), (i)   C (iv), (ii), (iii), (i)
B (iii), (ii), (iv), (i)   D (iii), (i), (iv), (ii)

45 Which one of the following blood vessels contain valves?
A veins   C capillaries
B arterioles   D arteries.

46   The purpose of a valve in the blood vessel in
Question **45** is to:
A propel blood along the vessel
B prevent backflow to the heart
C prevent bleeding from cut vessel
D prevent backflow to the body.

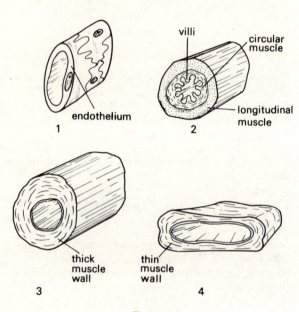

Figure 4.6

47   Which one of the four drawings in Figure 4.6 shows
the structure of an *artery*?

A 1   C 3
B 2   D 4.

48   Which one of the following blood vessels have thick
muscular walls and carry oxygenated blood?
A superior vena cava   C inferior vena cava
B pulmonary artery   D pulmonary veins.

49   The veins which contain most food as glucose and
amino acids are those from the:
A legs   C kidneys
B arms   D intestines.

50   The veins which contain the least amount of waste as
salts and urea are those from the:
A legs   C kidneys
B arms   D intestines.

51   In which one of the following veins is the blood
circulated through the liver tissue?
A systemic   C pulmonary
B hepatic portal   D renal.

52   Which one of the following organs has a vein leaving
it in addition to one vein and one artery supplying it?
A stomach   C kidney
B liver   D spleen.

53   Which two of the following vessels have valves?
(i) arteries, (ii) veins, (iii) lymphatics, (iv) capillaries.
A (i), (ii)   C (i), (iii)
B (iii), (iv)   D (ii), (iii)

54   Two of the following are true concerning arteries:
(i) they carry blood to the heart,
(ii) they have thin elastic walls,
(iii) arteries have narrow cavities,
(iv) the blood pressure is greater in arteries,
(v) arteries collapse easily when empty,
(vi) the pressure of surrounding skeletal muscles
      compress arteries.
A (i), (ii)   C (iv), (v)
B (iii), (iv)   D (iii), (vi)

55   Which one of the following arteries supplies the head?
A subclavian   C femoral
B carotid   D coronary.

56 Which four of the following are systemic veins?
(i) jugular, (ii) vena cava, (iii) portal, (iv) pulmonary,
(v) renal, (vi) hepatic.
A (i), (ii), (iii), (iv)   C (i), (ii), (v), (vi)
B (ii), (iii), (iv), (v)   D (iii), (iv), (v), (vi)

57 Fluids pass from the blood capillary network to tissues
and cells due to the influence of two of the following
pressures:
(i) atmospheric, (ii) osmotic, (iii) blood, (iv) spinal fluid.
A (i), (ii)   C (iii), (iv)
B (ii), (iii)   D (i), (iv)

58 Lymph is a fluid found:
A in intestinal juice,
B within lymphatic vessels,
C bathing cells,
D in blood clots.

59 Tissue fluid can return to the heart by two of the
following:
(i) lymphatic vessels, (ii) ureters, (iii) veins, (iv) oviducts.
A (i), (ii)   C (i), (iii)
B (ii), (iv)   D (iii), (iv)

60 One of the following is present in greater amounts in
lymph than in arterial plasma:
A lipids   C urea
B oxygen   D salts.

61 Which one of the following cells are present to a
considerable extent in lymph fluid?
A white blood cells   C platelets
B red blood cells   D erythrocytes.

62 The lacteal vessels of the small intestine discharge their
contents into one of the following systems:
A arterial   C lymphatic
B urinary   D venous.

63 Lymph fluid enters the circulatory system through
one of the following:
A pulmonary vein   C portal vein
B superior vena cava D aorta.

64 Which four of the following are functions of the lymph
glands or lymph nodes?
(i) insulin manufacture, (ii) removing bacteria,
(iii) red blood cell manufacture, (iv) making antibodies,
(v) intestinal juice secretion, (vi) white blood cell
manufacture.
A (i), (ii), (iii)   C (ii), (iv), (vi)
B (iv), (v), (vi)   D (i), (iii), (v)

65 Which one of the following glands become swollen and
painful when nearby tissues are sites of considerable
bacterial infection?
A sebaceous   C endocrine
B lymph   D sweat.

66 Which one of the following is a lymphatic gland?
A tonsil   C thyroid gland
B salivary gland D pancreas.

67 Which one of the following can destroy red blood cells
and also manufacture white blood cells?
A thymus gland   C red bone marrow
B lymph gland   D spleen.

68 Two of the following statements are correct:
(i) lymph glands filter blood, (ii) kidneys filter blood,
(iii) all arteries carry oxygenated blood, (iv) all veins
carry blood to the heart.
A (i), (ii)   C (ii), (iv)
B (i), (iii)   D (iii), (iv)

69   Which four of the following contain a fluid similar in composition to lymph?
(i) foot 'blisters', (ii) pericardium, (iii) inner ear, (iv) urinary bladder, (v) salivary gland, (vi) aqueous humour.

A (i), (ii), (iii), (vi)   C (ii), (iv), (v), (vi)
B (i), (ii), (iii), (iv)   D (ii), (iii), (iv), (vi)

70   Which three of the following body regions have most lymphatic glands?
(i) feet, (ii) armpits, (iii) hands, (iv) neck, (v) scalp, (vi) groins.

A (i), (ii), (iii)   C (i), (iii), (v)
B (iv), (v), (vi)   D (ii), (iv), (vi)

Questions 71–74 refer to Table 4.2.

TABLE 4.2

Organ	Material entering blood	Material leaving blood
Intestine	W	Oxygen
Kidneys	Carbon dioxide	X
Lungs	Y	Carbon dioxide
Ductless endocrine glands	Z	Oxygen and food

Use the following in your answers:
1 oxygen
2 urea and salts
3 hormones
4 food, amino acids, and glucose.

71   What is W?
A 1   C 3
B 2   D 4.

72   What is X?
A 1   C 3
B 2   D 4.

73   What is Y?
A 1   C 3
B 2   D 4.

74   What is Z?
A 1   C 3
B 2   D 4.

5

Respiration, Excretion, and the Skin

RESPIRATION

1   Which one of the following gases is needed for the aerobic respiration of foodstuffs in the body?
A carbon dioxide   C oxygen
B nitrogen   D inert gases.

2   Which one of the following gives the correct percentage composition of room or atmospheric air?

	Oxygen	Nitrogen	Carbon dioxide	Inert gases
A	20.0	0.3	78.0	1.0
B	0.3	20.0	78.0	1.0
C	1.0	78.0	0.03	20.0
D	20.0	78.0	0.03	1.0

3   The percentage composition of the air breathed out or exhaled by human lungs is:

	Oxygen	Nitrogen	Carbon dioxide
A	78.0	20.0	0.03
B	17.0	79.5	3.5
C	20.0	0.03	79.0
D	3.5	79.5	17.0

4   Which one of the following contains the greatest amount of carbon dioxide?
A venous blood   C room air
B arterial blood   D exhaled or expired air.

5   Which one of the following contains the least amount of oxygen?
A venous blood   C room air
B arterial blood   D exhaled or expired air.

Questions 6–11 refer to Figure 5.1.

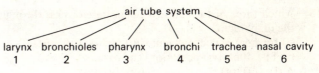

Figure 5.1

6   When air is inhaled it flows through the air tube system to reach the air sacs. What order does it follow?
A 6, 5, 4, 3, 2, 1   C 6, 3, 1, 5, 4, 2
B 6, 4, 1, 5, 3, 2   D 6, 4, 2, 5, 3, 1.

7   Which labelled part of the system contains the vocal cords?
A 1   C 3
B 2   D 4.

8   To which labelled part of the system is the epiglottis attached?
A 1   C 4
B 2   D 5.

9   Which labelled part of the system is common to the passage of both food and air?
A 6   C 5
B 3   D 1.

10 Which labelled part immediately warms, cleans, and
moistens the inhaled air?
A 3   C 5
B 4   D 6.

11 Which labelled parts of the system are contained
within the thorax?
A 1, 2, 3   C 3, 4, 5
B 2, 4, 5   D 1, 3, 6.

12 The intercostal muscles, vertebral column, diaphragm,
sternum, and ribs are all component parts of one of the
following:
A thorax   C pelvic cavity
B abdomen   D neck.

13 Which three of the following penetrate or pass through
the diaphragm?
(i) trachea, (ii) oesophagus, (iii) pleura, (iv) aorta,
(v) bile duct, (vi) posterior vena cava.
A (i), (iii), (v)   C (iv), (v), (vi)
B (ii), (iv), (vi)   D (i), (ii), (iii)

14 Which three of the following are contained *within* the
thorax?
(i) larynx, (ii) alveoli, (iii) thymus gland, (iv) mammary
glands, (v) spleen, (vi) pleura.
A (ii), (iii), (vi)   C (iv), (v), (vi)
B (i), (ii), (iii)   D (i), (iv), (v)

15 Which one of the following allows the lungs to move
freely and without friction between the inner walls of
the thorax?
A mucus within the bronchioles
B fluid within the pleura
C moisture in inhaled air
D pulmonary blood supply.

16 Which one of the following has walls one cell in thickness
and is well supplied with blood vessels?
A bronchi   C alveoli
B trachea   D bronchioles.

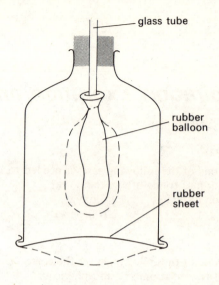

glass tube

rubber balloon

rubber sheet

Figure 5.2

17 The apparatus shown in Figure 5.2 demonstrates
ventilation movements of the lungs due to movement
of the:
A ribs   C diaphragm
B intercostal muscles   D sternum.

18 Three essential requirements for exchange of gases
through respiratory membranes are:
(i) moistness, (ii) dryness, (iii) of small surface area,
(iv) of great surface area, (v) many cells thick, (vi) one
cell thick.
A (i), (iv), (v)   C (i), (iv), (vi)
B (ii), (iii), (vi)   D (ii), (iv), (v)

19 Which one of the following are the respiratory organs
of insects?
A air sacs   C tracheoles
B bronchi   D skin.

20 Three of the following are used as respiratory organs during the complete life cycle of a frog:
(i) lungs, (ii) air sacs, (iii) skin, (iv) tracheoles, (v) gills, (vi) swim bladders.

A (i), (ii), (iv)   C (ii), (iii), (iv)
B (i), (iii), (v)   D (iii), (iv), (vi)

21 Which one of the following is the respiratory organ of an earthworm?

A air sacs   C skin
B gills   D lungs.

22 Single-celled animals (such as the amoeba) are without complex respiratory organs because:
A they can live without oxygen,
B they make their own oxygen,
C the cell membrane is permeable to oxygen,
D their ingested food already contains oxygen.

23 Which one of the following has the greatest body surface area to volume ratio useful for respiratory purposes?

A elephant   C shrimp
B amoeba   D rabbit.

24 The amount of air a healthy adult's lungs will hold is:

A 500 cm³   C 2.5 litres
B 5 litres   D 250 cm³.

25 During quiet or shallow breathing the volume of tidal air being breathed in and out is:

A 500 cm³   C 2.5 litres
B 5 litres   D 1 litre.

26 Which four of the following mechanisms are employed in *expiration* of lung air during very vigorous exercise?
(i) intercostal muscles contract, (ii) intercostal muscles relax, (iii) diaphragm muscle relax, (iv) diaphragm muscle contract, (v) abdominal muscles relax, (vi) abdominal muscles contract, (vii) thorax volume decreased, (viii) thorax volume increased.

A (i), (iv), (vi), (vii)   C (ii), (iv), (v), (viii)
B (ii), (iii), (vi), (vii)   D (i), (iii), (v), (viii)

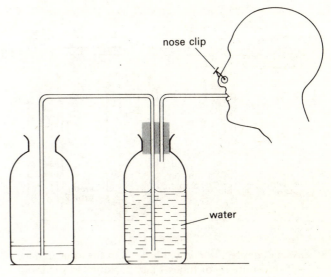

nose clip

water

Figure 5.3

27 The apparatus shown in Figure 5.3 could be used to:
A show expired air contains water vapour,
B show expired air contains carbon dioxide,
C measure the volume of air expired,
D show oxygen dissolves in water.

28 In which part of a mammal's respiratory system is the air stationary and not expelled during forced expiration?

A trachea   C bronchioles
B bronchi   D alveoli.

29 What is the name of the process by which gases in the lung air sacs are exchanged with gases leaving the blood vessels through the respiratory membranes?

A osmosis   C diffusion
B absorption   D filtration.

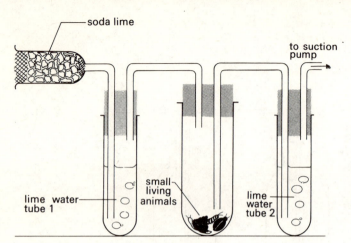

Figure 5.4

30   What changes, if any will be seen in the appearance of the contents of tubes 1 and 2, shown in Figure 5.4 when air is drawn through the apparatus containing small living animals?
A tube 1 only turns cloudy
B tube 2 only turns cloudy
C tubes both 1 and 2 turn cloudy
D tubes both 1 and 2 remain clear.

31   The experiment in Question 30 shows one of the products of breathing.
A oxygen   C heat
B carbon dioxide   D water vapour.

32   Which one of the following will have the greatest concentration or percentage composition of carbon dioxide?
A expired lung air   C alveolus air
B room air   D pulmonary artery blood.

33   Which two of the following control breathing?
(i) heart, (ii) brain, (iii) liver, (iv) blood carbon dioxide concentration, (v) blood sodium chloride concentration, (vi) blood urea concentration.
A (i), (v)   B (ii), (iv)   C (iii), (vi)   D (iv), (v)

34   Which one of the following is the principal carrier of oxygen from the lungs to the body cells?
A red blood cells   C platelets
B white blood cells   D plasma.

35   Carbon dioxide returns from the body cells to the lungs by means of two of the following:
(i) platelets, (ii) red blood cells, (iii) blood plasma, (iv) white blood cells.
A (i), (ii)   C (iii), (iv)
B (ii), (iii)   D (i), (iv)

36   In which one of the following chemicals is carbon dioxide transported to the lungs from the body tissues?
A urea   C sodium bicarbonate
B sodium carbonate   D carboxy-haemoglobin.

37   Select the best definition of respiration.
A transport of gases in the blood
B expiration and exhalation
C releasing energy within cells
D exchanging gases within the lungs.

38   Which one of the following situations cause more oxygen to pass into the blood:
A living at high altitudes
B flying in unpressurized aircraft at high altitude without breathing equipment
C working underground in compressed air tunnels
D living at sea level.

39   Which one of the following pure food components produces the greatest amount of energy from equal quantities?
A proteins   C carbohydrates
B lipids   D ethyl alcohol.

40   Which one of the following is the energy carrying chemical of a living cell?
A oxyhaemoglobin   C urea
B adenosine triphosphate   D sodium bicarbonate

41   Which one of the following is the site or place where respiration occurs inside a living cell?
A cell membrane   C mitochondria
B nucleus   D nuclear membrane.

42   In vigorous exercise (when the muscles are unable to receive oxygen) glucose is changed into one of the following to produce energy:
A lactic acid   C urea
B amino acids   D glycogen.

43   Which of the following can be respired or used to provide energy by means of aerobic respiration within cells?
(i) carbohydrates, (ii) minerals, (iii) vitamins, (iv) lipids, (v) water, (vi) proteins.
A (i) only   C (ii), (vi)
B (i), (iv), (vi)   D (ii), (iii), (v)

EXCRETION

1   Select the most suitable definition of excretion.
A removal of indigestible faeces
B liberation of energy from food
C removal of chemical wastes
D dissolving of insoluble foods.

2   Which four of the following function as excretory organs?
(i) lungs, (ii) spleen, (iii) kidneys, (iv) pancreas, (v) liver, (vi) skin.
A (i), (ii), (iv), (vi)   C (ii), (iii), (iv), (vi)
B (i), (iii), (v), (vi)   D (ii), (iii), (iv), (v)

3   One of the following is *not* a product of body cell metabolism.
A urea   C carbon dioxide
B faeces   D ammonia.

4   Which one of the following organs lie close to a fat depot?
A liver   C kidneys
B lungs   D pancreas.

5   Mammalian kidneys are located in the:
A thorax   C dorsal abdomen
B pelvic cavity   D ventral abdomen.

6   Which one of the following organs produces the colour pigment of urine as an excretory waste?
A kidneys   C liver
B lungs   D large intestine.

7   The urinary bladder of mammals is situated in the cavity of the:
A thorax   C pelvis
B ventral abdomen   D dorsal abdomen.

8   Which three of the following are excretory products of the skin?
(i) carbon dioxide, (ii) urea, (iii) water, (iv) sodium chloride, (v) sodium bicarbonate, (vi) urobilin.
A (i), (ii), (iii)   C (iii), (iv), (v)
B (ii), (iii), (iv)   D (iv), (v), (vi)

9   Which one of the following excretes surplus water from the contents of the alimentary canal?
A duodenum   C large intestine
B liver   D salivary glands.

10   What are the two excretory products of the lungs?
(i) carbon dioxide, (ii) oxygen, (iii) water, (iv) ammonia.
A (i), (ii)   C (iii), (iv)
B (i), (iii)   D (ii), (iv)

11   Which one of the following connects the kidneys with the urinary bladder?
A urethra   C ureters
B uterus   D oviducts.

12   Which one of the following connects the urinary bladder with the exterior?
A urethra   C ureters
B uterus   D oviducts.

13 The simple function of the kidneys is to act as:
A blood filters   C urine reservoirs
B blood pumps   D blood stores.

14 Urea is a metabolic product from the breakdown of:
A glucose   C proteins
B lipids   D alcohol.

15 Which of the following shows the correct order in which water is expelled from the body by the greatest to the least amount?
A urine − faeces − sweat − lung
B sweat − urine − faeces − lung
C urine − sweat − lung − faeces
D lung − urine − sweat − faeces.

16 Which two of the following show peristalsis?
(i) ureter, (ii) oesophagus, (iii) aorta, (iv) trachea.
A (i), (ii)   C (iii), (iv)
B (ii), (iii)   D (i), (iv)

Questions **17-19** refer to Figure 5.5.

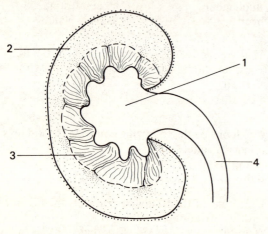

Figure 5.5

17 Which labelled part of the kidney is well supplied with blood vessels and shows a dark colour?
A 1   C 3
B 2   D 4

18 Where are the nephrons found?
A 4, 2   C 2, 3
B 1, 3   D 4, 1.

19 Where is the pelvis of the kidney?
A 1   C 3
B 2   D 4

Questions **20−22** refer to Figure 5.6 showing the structure of a single kidney tubule.

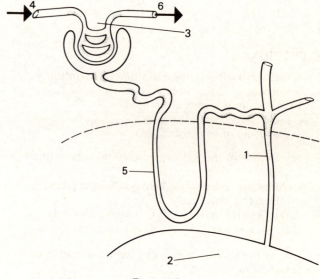

Figure 5.6

20 Where does the process of blood filtration under high pressure occur?
A 1   C 3
B 2   D 5

21   Where does selective reabsorption of amino acids, glucose, and salts from the kidney fluid occur?
A 1   C 4
B 2   D 5

22   The correct naming of the blood vessels reading from 4 to 6 by way of 3 would be:
A arteriole – glomerulus – venule
B venule – capillary – arteriole
C arteriole – glomerulus – arteriole
D venule – capillary – venule.

23   The process of urinating is called:
A parturition   C micturition
B defaecation   D ingestion.

24   Which two of the following have sphincter muscles?
(i) tongue, (ii) urethra, (iii) stomach, (iv) heart, (v) liver, (vi) trachea.
A (i), (ii)   C (iv), (v)
B (ii), (iii)   D (v), (vi)

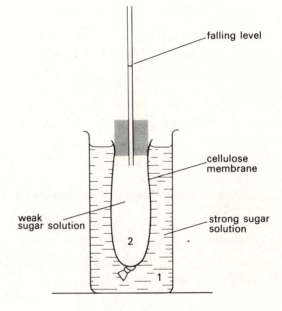

Figure 5.7

25   The steady control of the water and salt content of the body is through:
A thermostats   C humidostats
B homeostats   D pressurestats.

26   Which one of the following glands controls kidney function?
A thyroid   C pituitary
B adrenals   D thymus.

27   What is the cause of the falling level of weak sugar solution in the glass tube shown in Figure 5.7.

A water passes from 1 to 2
B water passes from 2 to 1
C sugar passes from 1 to 2
D sugar passes from 2 to 1.

28   In addition to water, which three of the following are normal components of urine?
(i) protein, (ii) pigment, (iii) glucose, (iv) urea, (v) salts, (vi) blood.
A (i), (ii), (iii)   C (ii), (iv), (v)
B (iv), (v), (vi)   D (i), (iii), (vi)

29   In which body organ is urea produced or manufactured?
A kidney   C urinary bladder
B pancreas   D liver.

30   Which one of the following blood vessels brings blood to the kidneys?
A renal vein   C renal artery
B hepatic artery   D pulmonary artery.

SKIN

1 Which layer of skin cells do not receive nourishment from the blood vessels and are therefore dead or dying?
A dermis C endoderm
B ectoderm D epidermis.

2 The horny protein material found in the outer layer of cells in the skin is called:
A glycogen C elastin
B keratin D sebum.

3 The layer of skin cells closest to the subcutaneous fat layer is called the:
A dermis C endodermis
B endoplasm D epidermis.

Questions 4 and 5 refer to Figure 5.8.

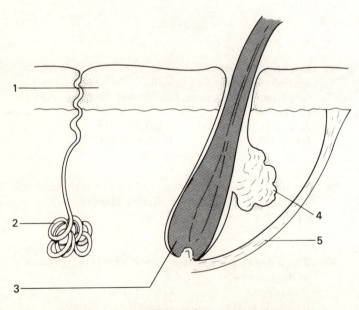

Figure 5.8

4 Which labelled part of the skin produces an oily or fatty secretion?
A 3 C 4
B 2 D 5

5 Which labelled part points to part of the mammalian excretory system?
A 3 C 4
B 2 D 5

6 The hairs of the skin are contained in narrow pits called the:
A pores C papillae
B follicles D ducts.

7 Which two of the following are the main component layers of the skin?
(i) subcutaneous fat, (ii) dermis, (iii) subcutaneous muscle, (iv) epidermis, (v) hair, (vi) cuticle.
A (i), (ii) C (iv), (vi)
B (ii), (iv) D (iii), (v)

8 Where in the human body is the skin thinnest?
A soles of the feet C the eye lids
B the abdomen D palms of the hands.

9 Where in Figure 5.8 is the unstriped smooth involuntary muscle found?
A 1 C 4
B 3 D 5

10 Which three of the following tissues are components of skin?
(i) striated voluntary muscle, (ii) bone, (iii) elastic fibres, (iv) nerves, (v) epithelia, (vi) cartilage.
A (i), (ii), (iii) C (ii), (iv), (vi)
B (iii), (iv), (v) D (i), (iii), (vi)

11 Which three of the following are glands of the skin?
(i) mammary, (ii) salivary, (iii) lymphatic, (iv) thyroid, (v) sebaceous, (vi) sweat.
A (iii), (iv), (v) C (i), (v), (vi)
B (ii), (iii), (iv) D (i), (ii), (vi)

12   Which one of the following vitamins can be manufactured in the skin?
A. vitamin C
B. vitamin A
C. vitamin B_{12}
D. vitamin D_3.

13   Which one of the following rays is needed to produce the vitamin manufactured in the skin?
A infra-red rays
B ultra-violet rays
C X-rays
D visible light rays.

14   If the vitamin produced by the skin is prevented from being formed, and is also absent from the diet, which disease will result by the deficiency?
A scurvy
B night blindness
C beri-beri
D rickets.

15   The skin is a very poor means of estimating:
A smoothness
B roughness
C temperature
D coldness.

16   Which four of the following is a means of gaining heat by the body?
(i) sweating, (ii) shivering, (iii) cold food, (iv) infra-red ray treatment, (v) cold showers or baths, (vi) vigorous exercise, (vii) hot food, (viii) iced drinks.
A (i), (iii), (v), (viii)
B (ii), (iv), (vi), (vii)
C (iii), (v), (vii), (viii)
D (iv), (v), (vi), (vii)

17   Which four of the following is a means of losing heat from the body?
(i) micturition, (ii) sweating, (iii) hot drinks, (iv) panting breath, (v) defaecation, (vi) hot baths, (vii) shivering (viii) capillary constriction.
A (i), (ii), (iv), (v)
B (i), (ii), (iii), (viii)
C (iii), (iv), (vi), (viii)
D (iv), (v), (vii), (viii)

18   Which four of the following functions in insulating the body heat?
(i) sweat glands, (ii) subcutaneous fat, (iii) erector pilae muscle contraction, (iv) capillary vessel contraction, (v) interlocking hairs, (vi) sebaceous glands.
A (i), (ii), (iii), (iv)
B (ii), (iii), (iv), (v)
C (iii), (iv), (v), (vi)
D (i), (iii), (v), (vi)

19   Which one of the naked animals in Table 5.1 will lose heat at the most rapid rate in cold surroundings?

TABLE 5.1

Animal	Surface area of skin (cm^2)	Total body volume (cm^3)
1	350	100
2	800	400
3	600	100
4	150	100

A 1
B 2
C 3
D 4

20   The maintenance of a constant body temperature, irrespective of the temperature of the surroundings, is found in:
(i) fish, (ii) amphibia, (iii) mammals, (iv) shellfish, (v) birds, (vi) reptiles.
A (i), (ii)
B (ii), (iii)
C (iii), (v)
D (iv), (vi)

21   The animals with the ability described in Question **20** are called:
A homeostatic
B poikilothermic
C thermonastic
D homoiothermic.

22   Blushing or flushing of the skin is due to:
A capillary vessel dilatation
B capillary vessel constriction
C capillary vessel bleeding
D pigment cell dilatation.

23   Fibres, such as hairs and feathers, prevent heat loss from animal bodies by means of the:
A fibres matting together
B sweat clinging to the fibres
C grease coating the fibres
D air between the fibres.

24   In which one of the following environments do animals encounter the least sudden change of temperature?
A moorlands   C mountains
B sea water   D meadows.

25   Which three of the following skin components function to prevent water loss from the body and also water-proof the skin?
(i) sebaceous glands, (ii) hairs, (iii) dermis, (iv) horny epidermis, (v) sweat glands, (vi) subcutaneous fat.
A (i), (ii), (iv)   C (ii), (iv), (vi)
B (iii), (v), (vi)   D (i), (iii), (v)

26   What is the natural sun-screen material of human skin affording protection against harmful sun rays?
A sebum secretion   C hair fibres
B sweat secretion   D dermis pigment.

27   Which one of the following lubricates the skin and keeps it supple?
A sebum   C mucus
B sweat   D saliva.

28   The natural antiseptic material of the skin which helps to fight skin microbes is:
A sebum   C mucus
B sweat   D lymph.

29   What is the possible function of cats and rabbits whiskers or vibrissae?
A message receptors   C sweat controls
B feeler gauges   D food samplers.

30   Which four of the following mechanisms will restore the body temperature to normal for a person who has been immersed in icy water?
(i) increased sweating, (ii) increased metabolism,
(iii) decreased sweating, (iv) capillary vasoconstriction,
(v) capillary vasodilatation, (vi) decreased metabolism,
(vii) shivering, (viii) panting breath.
A (i), (ii), (iii), (iv)   C (v), (vi), (vii), (viii)
B (ii), (iii), (iv), (vii)   D (i), (v), (vi), (viii)

31   Which three of the following control the water content of the animal body?
(i) stomach, (ii) large intestine, (iii) heart, (iv) liver,
(v) kidneys, (vi) lungs.
A (i), (ii), (iii)   C (ii), (v), (vi)
B (iii), (iv), (v)   D (i), (iv), (vi)

6

Sensory Organs and Endocrine Systems

1 Which three of the following systems unite or co-ordinate the activities of all the living cells of the body?
(i) skeleton, (ii) alimentary, (iii) nervous, (iv) reproductive, (v) endocrine, (vi) circulatory.
A (i), (ii), (iii)   C (iii), (v), (vi)
B (iv), (v), (vi)   D (i), (ii), (iv)

2 The system informing the mammalian body of changes in its internal and external environment is called the:
A nervous   C circulatory
B reproductive   D endocrine.

3 Which four of the following are detected through the sensory receptors of the human body?
(i) heat, (ii) ultra-violet rays, (iii) pressure, (iv) sound, (v) radio waves, (vi) chemicals, (vii) radar, (viii) X-rays.
A (i), (ii), (vii), (viii)   C (ii), (v), (vii), (viii)
B (i), (iii), (iv), (vi)   D (iii), (iv), (v), (vi)

4 Which four of the following are stimuli detected through the *scattered* sensory receptors in contrast to the *localized* special sense organs?
(i) gravity, (ii) light, (iii) chemicals, (iv) cold, (v) pressure, (vi) pain, (vii) heat, (viii) sound.
A (i), (ii), (iii), (iv)   C (iv), (v), (vi), (vii)
B (ii), (iii), (iv), (v)   D (v), (vi), (vii), (viii)

5 Proprioreceptors are found in one of the following:
A skin   C muscles
B nose   D eye.

6 Which one of the following possess exteroreceptors?
A skin·   C tendons
B muscles   D heart.

7 Which one of the following is without sensory cells or nerves?
A teeth   C skin
B nails   D bone.

8 An erect posture is maintained by four of the following:
(i) eyes, (ii) nose, (iii) ears, (iv) proprioreceptors, (v) tongue, (vi) skin.
A (i), (iii), (iv), (vi)   C (i), (ii), (iii), (v)
B (ii), (iv), (v), (vi)   D (iii), (iv), (v), (vi)

Select your answers to Questions 9-12 from the following:
(i) cranial nerves, (ii) ears, (iii) spinal nerves, (iv) spinal cord, (v) brain, (vi) striated muscles, (vii) sweat glands, (viii) proprioreceptors.

9 Which two are typical sense organs?
A (i), (iv)   C (ii), (vi)
B (ii), (viii)   D (iv), (v)

10 The central nervous system consists of two main parts:
A (i), (iii)   C (iv), (v)
B (iii), (iv)   D (i), (v)

11 The peripheral nervous system consists of two parts:
A (i), (iii)   C (iv), (vi)
B (ii), (viii)   D (v), (vii)

12   Which one is an example of an effector organ?
A (ii)   C (iv)
B (vi)   D (viii)

13   Receptor ⌐
              Intermediate or
              connector neurone
      Effector ⌐

The above is a summary of a reflex action. In which
two of the following are connector or intermediate
neurones found?
(i) spinal nerve, (ii) spinal cord, (iii) cranial nerve,
(iv) brain, (v) synapse, (vi) dendrites.
A (i), (iii)   C (iii), (vi)
B (ii), (iv)   D (iv), (v)

14   The small gap through which connection is made
between adjacent nerve cell bodies is called the:
A dendrite   C synapse
B axon   D end plate.

15   Which one of the following is an effector organ of
the eye?
A lens   C retina layer
B ciliary muscle   D cornea layer.

Questions 16–18 refer to Figure 6.1 illustrating a motor
neurone.

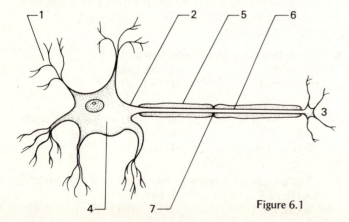

Figure 6.1

16   Which one of the following show the correct path of
nervous impulse flow through the motor neurone?
A 1 → 2 ← 3   C 1 → 2 → 3
B 1 ← 2 ← 3   D 1 ← 2 → 3.

17   Which labelled part of the neurone structure forms the
grey matter?
A 4   C 6
B 7   D 5

18   What is the name given to part labelled 6 that conducts
impulses through the neurone?
A dendron   C cell body
B sheath   D axon.

19   Which one of the following is provided with a cranial
nerve?
A arm   C eye
B leg   D trunk.

20   Which one of the following has a spinal nerve?
A skin of arm   C eye retina
B ear cochlea   D tongue taste buds.

21   Select three functions of the autonomic nervous system
from the following:
(i) running, (ii) voice production, (iii) peristalsis, (iv) body
balance, (v) blushing, (vi) gland control.
A (i), (ii), (iii)   C (iii), (v), (vi)
B (i), (ii), (iv)   D (iv), (v), (vi)

22   Which three of the following are voluntary actions?
(i) pupil dilatation, (ii) heart beat, (iii) goose pimples,
(iv) speech, (v) walking, (vi) chewing.
A (i), (ii), (iii)   C (ii), (iv), (vi)
B (iv), (v), (vi)   D (i), (iii), (v)

23   Stretch receptors are found at two of the following.
(i) muscles, (ii) nose, (iii) tendons, (iv) eye, (v) bone,
(vi) skin.
A (i), (iii)   C (iii), (v)
B (ii), (iv)   D (v), (vi)

24 The purpose of the stretch receptors is for:
A posture and muscle movement
B bladder control
C elasticity of the skin
D expansion and elasticity of the lung.

25 The long nerve fibre of a sensory neurone is called the:
A mixed nerve   C ganglion
B dendron   D intermediate neurone.

26 Which one is the correct path of travel for an impulse
from the *receptor* to the *effector* in a reflex action?
A axon–sensory neurone–intermediate neurone–motor
neurone–axon
B axon–motor neurone–intermediate neurone–sensory
neurone–dendron
C dendron–sensory neurone–intermediate neurone–
motor neurone–axon
D dendron–motor neurone–intermediate neurone–
sensory neurone–axon.

Questions 27–29 refer to Figure 6.2.

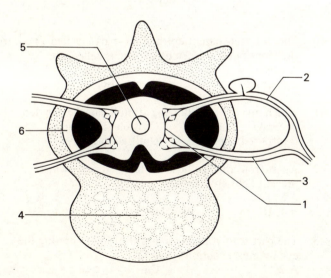

Figure 6.2

27 Which one of the following labelled parts indicate the
sensory nerve?
A 4   C 2
B 1   D 3

28 The fluid within and around the spinal cord parts labelled
5 and 6 is called:
A lymphatic   C plasma
B cerebrospinal   D pleural.

29 Part labelled 4 is called the:
A neural spine   C transverse process
B centrum   D neural arch.

30 The largest and most well-developed part of the human
brain is the:
A medulla   C cerebellum
B cerebrum   D mid-brain.

31 The medulla of the brain is concerned with amongst
other functions:
A balance and muscle co-ordination
B movement of body parts
C heart beat and breathing
D hearing and sight.

32 Which one of the following have association centres?
A autonomic nerves   C brain
B spinal cord   D spinal nerves.

33 Which one of the following animals is without a central
nervous system?
A earthworm   C bird
B hydra   D frog.

34 Which one of the following is most beneficial to the
nervous system of man?
A emotional shock   C sleep
B alcohol   D exercise.

35 Which one of the following is a *conditioned* reflex action?
A tapping below the knee causes knee jerk
B pupil contraction in bright light
C salivation on seeing succulent foods
D sneezing in a dusty room.

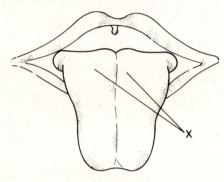

Figure 6.3

36 What type of taste sensation is experienced at position X shown in Figure 6.3?
A saltness   C sweetness
B sourness   D bitterness.

37 Sensory impulses from the taste buds travel to the central nervous system by way of the:
A cranial nerves   C autonomic nerves
B spinal nerves   D salivary glands.

38 The organ through which we appreciate food flavour is called the:
A olfactory   C auditory
B optic   D tactile.

Figure 6.4 showing the eye's structure is referred to in Questions **39–50**.

39 The component parts of the eye which serve to refract the light rays onto part labelled 8 of the eye are:
A 1, 6, 10, 12   C 4, 6, 7, 13
B 4, 7, 10, 13   D 2, 3, 7, 14.

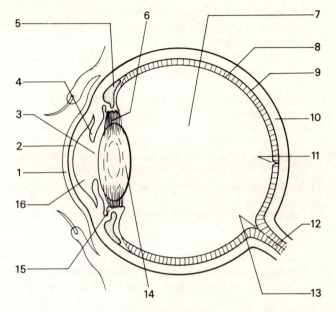

Figure 6.4

40 The thin transparent epithelium (part labelled 1) covering the front outer surface of the eye is called the:
A cornea   C conjunctiva
B choroid   D sclerotic.

41 Which one of the following stimuli can the eye detect?
A X-rays   C visible light rays
B infra-red rays   D ultra-violet rays.

42 Apart from the eye lids, the labelled parts which prevent light entering the eye are:
A 8, 14   C 2, 7
B 4, 10   D 1, 6

43 The part regulating the amount of light entering the eye by muscle action is labelled:
A 2   C 6
B 4   D 9

44 The process of accommodation involves which of the following labelled parts:

A 5, 6, 14, 15
B 2, 3, 4, 16
C 3, 4, 7, 14
D 8, 9, 11, 12.

45 Which one of the following can alter the shape of part labelled 14?

A eye lid sphincter muscle
B iris muscles
C eye orbit muscles
D ciliary muscles.

46 Which one of the following is the white of the eye?

A cornea
B sclerotic
C iris
D pupil.

47 The light sensitive layer of the eye is called the:

A vitreous humour
B choroid
C sclera
D retina.

48 The light receptors or cones of the eye have three of the following functions:
(i) act in dim light, (ii) act in bright light, (iii) see black and white only, (iv) see colour, (v) of great visual accuracy, (vi) poor visual accuracy.

A (i), (iii), (vi)
B (i), (iv), (v)
C (ii), (iii), (v)
D (ii), (iv), (vi)

49 The point or area of greatest visual accuracy within the eye for resolving fine detail is at part labelled:

A 8
B 11
C 12
D 13.

50 The blind spot is located at part labelled:

A 7
B 9
C 11
D 13.

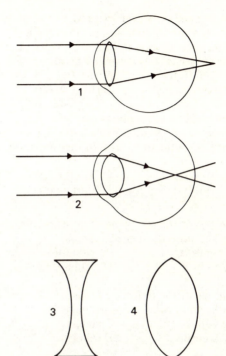

Figure 6.5

51 Which diagram in Figure 6.5 shows the disorder of the eye called short sight, and which lens will help to correct it?

A 1, 3
B 1, 4
C 2, 3
D 2, 4.

52 Shortsightedness is due to:

A a short eyeball
B a long eyeball
C a hardened lens
D unequal curvatures of cornea.

53 Stereoscopic vision allows a human to:
A estimate light intensity
B estimate depth and distance
C give an uninterrupted view around the head
D distinguish between red and green colours.

54 The tear gland secretion serves to:
(i) moisten the conjunctiva, (ii) feed the lens,
(iii) lubricate the eye, (iv) act as an antiseptic,
(v) make sebum, (vi) neutralize alkali.
A (i), (iii), (iv)   C (iii), (iv), (v)
B (ii), (v), (vi)   D (iv), (v), (vi)

55 The choroid layer functions to:
(i) reflect light, (ii) alter eyeball shape, (iii) refract light,
(iv) prevent reflection, (v) feed the retina, (vi) absorb light.
A (i), (ii), (iii)   C (ii), (iv), (vi)
B (iv), (v), (vi)   D (i), (iii), (v)

56 Which one of the following vitamins promotes night vision and prevents night blindness?
A. vitamin A   C. vitamin C
B. vitamin B_{12} D. vitamin D.

57 Which one of the following foodstuffs are good sources of the vitamin which prevents night blindness?
A egg yolk   C oranges
B black currants   D cabbage.

58 What are the three functions of the aqueous humour of the eye?
(i) feed the living lens, (ii) colour the pupil, (iii) feed the cornea, (iv) lubricate the conjunctiva, (v) refract light rays, (vi) whiten the eye.
A (i), (ii), (iii)   C (ii), (iv), (vi)
B (iv), (v), (vi)   D (i), (iii), (v)

59 Which three of the following animals have colour vision?
(i) bats, (ii) cats, (iii) monkeys, (iv) birds, (v) dogs, (vi) apes.
A (i), (ii), (v)   C (ii), (iii), (v)
B (i), (iv), (vi)   D (iii), (iv), (vi)

60 The number of vibrations of sound waves taking place in one second is called the:
A oscillation   C wavelength
B frequency   D velocity.

61 The unit for measuring sound vibrations per second is called the:
A joule   C Celsius
B hertz   D newton.

62 The vocal ligaments producing a high-pitched singing note are:
A long, slack, and thick   C short, tight and thin
B short, slack and thin   D long, tight, and thick.

Questions 64–77 refer partly to Figure 6.6 showing the *structure* of the ear and also the *function* of the ear in animals.

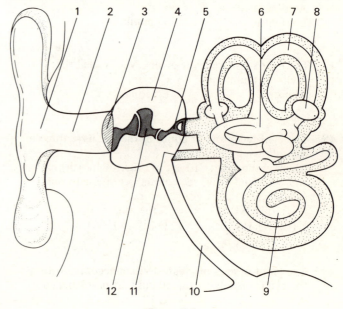

Figure 6.6

63 Which labelled part of the diagram is called the middle ear?
A 1   C 4
B 2   D 6.

64   Which of the following animals possesses an outer or
     external ear?
     A snakes   C frogs
     B moles   D dogfish.

65   Which one of the following divides the middle ear from
     the outer or external ear?
     A tympanum   C pinna
     B round window   D oval window.

66   Which labelled part of the ear contains small hairs, and
     produces cerumen?
     A 2   C 6
     B 4   D 10.

67   The labelled part of the ear containing the organ of
     hearing or sound sensory receptors is:
     A 3   C 9
     B 6   D 12.

68   The labelled parts of the ear concerned with the
     transmission of sound waves or vibrations are:
     A 3, 12   C 7,8
     B 10, 11   D 9, 10.

69   The organ of balance of the ear is located in parts
     labelled:
     A 3, 5   C 9, 10
     B 7, 8   D 10, 11.

70   Which part is labelled the ear ossicles?
     A 2   C 9
     B 6   D 12.

71   Air pressure in the middle ear is controlled by air
     entering and leaving through the:
     A external ear   C oval window
     B eustachian tube   D round window.

72   The stirrup bone or stapes connects with the:
     A round window   C ear drum
     B oval window   D eustachian tube.

73   Which labelled parts of the ear contain
     proprioreceptors?
     A 2, 3   C 7, 8
     B 4, 5   D 9, 10.

74   Which of the following is the correct path of sound
     transmission through the ear?
     A pinna – ossicles – tympanum – round window –
       vestibule
     B ossicles – pinna – tympanum – oval window –
       cochlea
     C pinna – tympanum – ossicles – oval window –
       cochlea
     D tympanum – ossicles – pinna – round window –
       vestibule.

75   Sound waves which have entered the inner ear escape
     by way of the:
     A oval window   C eustachian tube
     B round window   D external auditory meatus.

76   Which two of the following are fluids of the inner ear?
     (i) plasma, (ii) perilymph, (iii) mucus, (iv) cerebrospinal
     fluid, (v) endolymph, (vi) serum.
     A (i), (iii)   C (iii), (vi)
     B (ii), (v)   D (iv), (vi)

77   Endocrine glands co-ordinate all cell activity by producing
     hormones distributed through the body by the:
     A nerve fibres   C blood system
     B tubular ducts   D skeleton bone.

78   Which one of the following endocrine glands is closely
     associated with the function and is also part of the brian?
     A adrenal   C pituitary
     B parathyroid   D thymus.

79   Which endocrine gland secretion contains the element
     iodine?
     A thyroid   C pituitary
     B thymus   D adrenal.

80   The hormone insulin is produced in the:
    A pancreas   C liver
    B kidney   D adrenal gland.

81   When the human body is subjected to severe fright which
    gland is active during this period?
    A pancreas   C thyroid
    B adrenal   D spleen.

82   Which one of the following can change stored glycogen
    into glucose?
    A insulin   C thyroxine
    B adrenaline   D amylase.

83   Which one of the following glands can change glucose
    into glycogen and also help in the respiration of
    glucose by living cells?
    A pancreas   C thyroid
    B adrenals   D thymus.

Reproduction, Growth, and Heredity

Questions **1–6** refer to Figure 7.1 of the male mammal reproductive system.

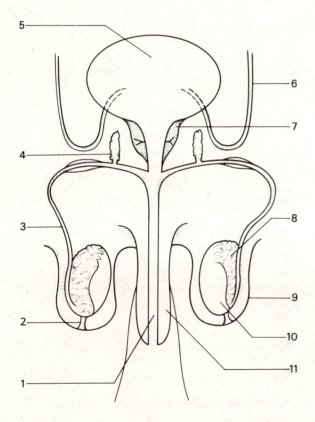

Figure 7.1

1 Where are the male gametes produced?
A 4   C 8
B 7   D 10.

2 The male gonads are labelled part:
A 4   C 10
B 7   D 11.

3 In the adult mammal the testes are situated within the:
A abdomen   C seminal vesicles
B scrotum   D urethra.

4 The production of male sperm is favoured by:
A steady temperatures above blood heat
B steady temperatures below blood heat
C normal blood heat
D fluctuating temperature changes above and below
    blood heat.

5 The part of the male reproductive system which conducts both urine and sperm is called the:
A ureter   C epididymis
B urethra   D vas deferens.

6 The labelled part producing the male sex hormone is:
A 4   C 8
B 7   D 10.

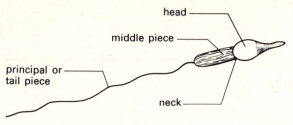

Figure 7.2

7 Which part of the male sperm, shown in Figure 7.2, contains the haploid nucleus?
A head
B neck
C middle piece
D principal or tail piece.

8 Which one of the labelled parts in Figure 7.1 are for the storage of sperm?
A 7
B 4
C 9
D 5.

9 The sperm are transported in a fluid produced by three of the following:
(i) urethra, (ii) vas deferens, (iii) seminal vesicles, (iv) prostate gland, (v) bladder, (vi) penis.
A (i), (ii), (iii)
B (ii), (iii), (iv)
C (iv), (v), (vi)
D (i), (v), (vi)

10 Three secondary characters which appear in a male at puberty are:
(i) larynx enlargement, (ii) breast growth, (iii) beard growth, (iv) hip enlargement, (v) skin softening, (vi) muscle development.
A (i), (iii), (vi)
B (ii), (iv), (vi)
C (i), (ii), (iv)
D (ii), (iii), (v)

Questions 11–14 refer to Figure 7.3 which shows the female mammal reproductive system.

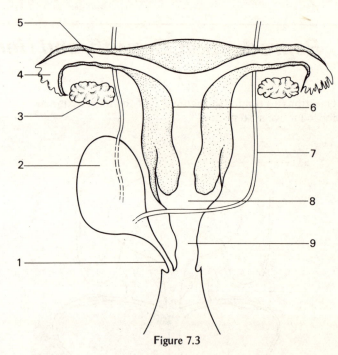

Figure 7.3

11 Graafian follicles are produced in part labelled:
A 3
B 4
C 5
D 6.

12 The female gonads are parts labelled:
A 3
B 4
C 5
D 6.

13 The female urethra is part labelled:
A 1
B 5
C 7
D 9.

14 The uterine fallopian tubes are parts labelled:
A 1
B 5
C 7
D 9.

15 External fertilization of the eggs takes place in three of the following:
(i) rat, (ii) herring, (iii) sparrow, (iv) housefly, (v) frog, (vi) trout.
A (i), (ii), (iii)   C (ii), (v), (vi)
B (i), (iv), (v)   D (iii), (iv), (vi)

16 Which one of the following is the male sex hormone?
A testosterone   C progesterone
B oestrogen   D adrenaline.

17 Which one of the following glands control the development of the sex organs?
A adrenal   C thyroid
B pituitary   D pancreas.

18 Oestrogen is produced by one of the following:
A testes   C corpus luteum
B Graafian follicles   D thymus gland.

19 Which one of the following encloses the female gamete in a fluid-filled sac?
A corpus luteum   C amnion
B Graafian follicle   D ovary.

20 The process of releasing the ripe female gamete from the ovary is called:
A parturition   C fertilization
B ovulation   D implantation.

21 The lining of the female reproductive system (shown in Figure 7.3) which is removed and periodically discharged together with a certain amount of blood is labelled part:
A 2   C 5
B 4   D 6.

22 In which labelled part of Figure 7.3 is the human zygote normally first formed?
A 3   C 6
B 5   D 9.

23 Which two of the following are sex hormones produced by the human ovary?
(i) testosterone, (ii) adrenaline, (iii) insulin, (iv) secretin, (v) oestrogen, (vi) progesterone.
A (i), (ii)   C (iv), (v)
B (ii), (iii)   D (v), (vi)

24 The endometrium is the lining of the:
A bladder   C uterus
B vagina   D oviduct.

25 Where does the fertilized egg normally undergo final growth and development?
A oviduct   C vagina
B uterus   D urethra.

26 The normal period of gestation in humans is:
A four weeks   C 5 months
B two months   D 9 months.

27 Which one of the following animals is hermaphrodite?
A cockroach   C frog
B earthworm   D sparrow.

28 The formation of offspring identical in all respects to the *single* parent producing them is seen in:
A identical twins   C fraternal twins
B asexual reproduction   D sexual reproduction.

29 Which three of the following can be reproduced vegetatively?
(i) apple, (ii) earthworm, (iii) strawberry, (iv) potato, (v) protozoa, (vi) mouse.
A (i), (ii), (iii)   C (ii), (iv), (v)
B (i), (iii), (iv)   D (iii), (iv), (v)

30   The implantation of the human embryo takes place in the:

A vagina   C oviduct
B uterus   D bladder.

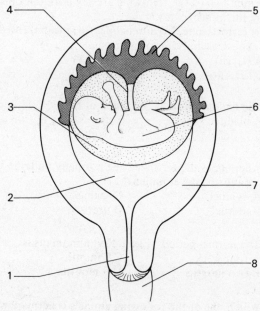

Figure 7.4

31   The disc of tissue connecting the embryo to the mothers womb shown in Figure 7.4 is called the:

A umbilicus   C placenta
B amnion   D chorion.

32   The amniotic fluid is found in which labelled part of Figure 7.4?

A 2   C 5
B 3   D 8.

33   The cord consisting of arteries and a vein connecting the mammalian embryo to the mother is called the:

A amnion   C umbilicus
B placenta   D chalaza.

34   The link between the blood of the embryo and the blood of the mother is:

A directly into the mothers blood vessels
B directly into the mothers heart
C by diffusion between blood vessels
D through the lymphatic vessels.

35   The embryo chick receives its oxygen supply by way of the:

A egg yolk   C placenta
B shell   D amniotic fluid.

36   Excretory products of the mammalian embryo are removed by the:

A amniotic fluid   C ureter
B filtration into   D placenta.
  vagina

37   The female sex hormones, produced by the mother in large amounts during pregnancy, are formed by the:

A Graafian follicles   C uterus
B placenta   D mammary glands.

38   Respiration of the mammalian embryo during pregnancy is:

A by oxygen diffusion through the placenta
B anaerobically without oxygen
C by diffusion of oxygen from amniotic fluid
D by diffusion of oxygen through the uterus wall.

39   The process of giving birth to the mammalian embryo is called:

A gestation   C parturition
B expiration   D menstruation.

40   During birth the embryo passes to the exterior by way of the:

A fallopian tube   C vagina
B oviduct   D urethra.

41 During pregnancy the developing embryo is given protection by the mother from one of the following, which may be circulating in the mothers blood:
A all drugs   C most bacteria
B german measles virus   D sexually transmitted diseases.

42 Which of the following young animals obtain their first food by suckling milk?
A fish   C birds
B amphibia   D mammals.

43 Which two of the following are features of the development of a frog?
(i) controlled environment, (ii) protection within mother, (iii) parental care, (iv) changing environment, (v) little parental care, (vi) period of weaning.
A (i), (ii)   C (iv), (v)
B (iii), (iv)   D (v), (vi)

44 Which of the following will produce quadruplets composed of a brother and sister and two identical twin sisters?
A four separate eggs giving a total of four embryos
B one egg having divided into four embryos
C two eggs giving two embryos and one egg giving two embryos
D two eggs each dividing into two and producing a total of four embryos.

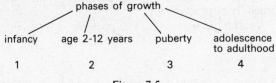

Figure 7.5

45 Which two of the growth phases (Figure 7.5) show the greatest growth rate?
A 1, 2   C 2, 3
B 1, 3   D 3, 4.

46 At what period in human growth is the head larger in relation to the total body size?
A birth   C 15 years
B 5 years   D 25 years.

47 Which one of the following is in the correct order showing the cycle of sexual reproduction?
A adult – fertilization – gametes – embryo – zygote
B fertilization – adult – gametes – zygote – embryo
C adult – gametes – fertilization – zygotes – embryo
D embryo – adult – zygote – gametes – fertilization.

Questions 48–57 refer to Figure 7.6, which is a diagram of a cell.

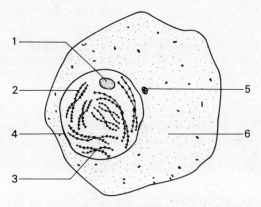

Figure 7.6

48 Which labelled part represents the nucleus of the cell?
A 1   C 5
B 4   D 6.

49 What is the structure part 3 called?
A centrosome   C mitochondria
B nucleolus   D chromosome.

50 The swellings or beads on part labelled 2 are called:
A centrosomes   C genes
B chromosomes   D nucleoli.

51   Where is the chemical deoxyribonucleic acid or DNA
     produced?
     A 1   C 5
     B 2   D 6.

52   What is the chromosome number of the cell nucleus?
     A 2   C 8
     B 4   D 12.

53   If the cell shown in Figure 7.6 is a normal body cell
     taken from the lining of the mouth, by which one of
     the following processes will its nucleus divide?
     A meiosis   C gametogenesis
     B mitosis   D reduction division.

54   If the cell is gamete-producing, taken from a testes, by
     what method would its nucleus divide to produce sperm?
     A meiosis   C fragmentation
     B mitosis   D binary fission.

55   What will the chromosome number be of the sperm cells
     produced by the nuclear division in Question 54?
     A 2   C 6
     B 4   D 12.

56   How many sperm cells will be produced by the division
     of the one gamete producing cell in Question 54?
     A 2   C 6
     B 4   D 12.

57   What is the diploid number of chromosomes in a normal
     human body cell nucleus?
     A 6   C 23
     B 12   D 46.

58   How are the chromosomes normally arranged in the
     resting body cells?
     A in pairs   C in fours
     B singly   D in threes.

59   How many chromosomes are there in the nucleus of a
     human ova or sperm?
     A 6   C 23
     B 12   D 46.

60   How many chromosomes are there in the nucleus of a
     fertilized ovum?
     A 23   C 69
     B 46   D 92.

61   If there are 40 chromosomes in the body cell nuclei of
     a mouse how many of these are *autosomes*?
     A 40   C 20
     B 38   D 18.

62   In a male human body cell nucleus there are two
     chromosomes given the identifying letters of:
     A XX   C XY
     B WX   D YZ.

63   What are the two chromosomes in Question 62 called?
     A centrosomes   C sex chromosomes
     B autosomes   D genes.

64   The effect of genes in body chemistry is directed
     mainly at producing:
     A carbohydrates   C proteins
     B lipids   D minerals.

65   Which one of the following is due to the body cell
     nuclei having more than the normal number of
     chromosomes?
     A haemophilia   C Down's syndrome or
                                mongolism
     B sickle-cell anaemia   D curly hair.

Questions **66–68** refer to Figure 7.7. Capital letters refer to the dominant gene.

B = blue eyes;   b = brown eyes;

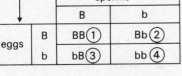

Figure 7.7

66 Which of the second filial generation are blue-eyed phenotype?
A 1
B 1, 3, 4
C 1, 2, 3
D 4.

67 Which one of the second filial generation is a homozygous blue-eyed genotype?
A 1
B 2
C 3
D 4.

68 Which one of the second filial generation will be a homozygous brown-eyed phenotype?
A 1
B 2
C 3
D 4.

Question **69** refers to Figure 7.8. Capital letters refer to the dominant gene.

69 Which of the following second filial generation will be the dominant homozygous phenotype produced by intermarriage of the first filial generation?
A wW
B ww
C Ww
D WW.

W = wavy hair;   w = straight hair

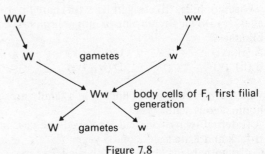

Figure 7.8

Question **70** refers to Figure 7.9.

R = right handed   r = left handed

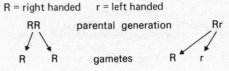

Figure 7.9

70 What percentage of the first filial generation will be right-handed children?
A 25
B 50
C 75
D 100.

71 Which three of the following are known to cause mutations in human offspring?
(i) atomic radiation, (ii) sound waves, (iii) certain drugs, (iv) radio waves, (v) X-rays, (vi) sunlight.
A (i), (ii), (iv)
B (i), (iii), (v)
C (ii), (iv), (v)
D (iii), (v), (vi)

72　Two of the following are inherited through sex linkage.
(i) haemophilia, (ii) straight hair, (iii) hare lip, (iv) blue
eyes, (v) Down's syndrome or mongolism, (vi) colour
blindness.
A (i), (vi)　　　　　　C (iii), (iv)
B (ii), (v)　　　　　　D (iv), (vi)

73　Which of the following children in a family are
homozygous genotypes?
A fraternal twins Tom and Jane aged one year
B Jack and Kate aged 3 and 5 years
C dizygotic twins Mary and Bob aged 7 years
D monozygotic twins Tim and Jim aged 9 years.

74

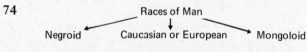

Figure 7.10

Select the four features typical of the Mongoloid race
from Table 7.1.

TABLE 7.1

Skin	Hair	Nose	Other
1 fair	4 wavy/straight	7 small flat	10 broad lips
2 dark	5 black curly	8 broad flat	11 high cheek bones
3 yellow or red	6 dark straight	9 narrow	12 prominent brows

A 1, 4, 9, 10　　　　C 3, 6, 7, 11
B 2, 4, 9, 12　　　　D 2, 5, 8, 10.

75　Which of the following is the more satisfactory method
for classifying the races of man?
A hair colour and　　C blood groups
　texture
B nose shape　　　　D skin colour.

Fish, Amphibia, Birds, and Mammals

1 Scales are a distinguishing feature of:
A fish   C amphibia
B birds   D mammals.

2 Where are otoliths found in the vertebrate body?
A eye   C ear
B kidney   D long bones.

3 The approximate age of a fish can be estimated by the growth rings appearing in the:
(i) skin scales, (ii) muscle segments, (iii) long bones, (iv) eye lens, (v) otoliths, (vi) fins.
A (i), (v)   C (iii), (iv)
B (ii), (iv)   D (iv), (vi)

4 A lateral line sense organ in a fish serves to detect:
A temperature changes   C sound waves
B water vibrations   D salinity changes.

5 The gill covering in fish is called the:
A spiracle   C pleura
B operculum   D thorax.

6 Gills of vertebrates can be slits in the:
A oesophagus   C pharynx
B trachea   D nasal cavity.

7 Water is brought to the gills of a fish by way of the:
A nostrils   C swim bladder
B mouth   D gill slits.

8 The fine blood capillary network of gills is located in the:
A rakers   C filament
B operculum   D pouch.

9 Segmentation of the vertebrate body of a fish is clearly seen in the:
A paired fins   C heart
B alimentary canal   D body muscles.

10 Which two of the following fins brakes or stops a moving fish?
(i) caudal, (ii) dorsal, (iii) ventral, (iv) pelvic, (v) anal, (vi) pectoral.
A (i), (ii)   C (iv), (v)
B (ii), (iii)   D (iv), (vi)

11 Which two of the fins in Question 10 are paired fins?
A (i), (ii)   C (iii), (v)
B (ii), (iv)   D (iv), (vi)

12 Which one of the following is a bony fish?
A shark   C cod
B dogfish   D skate or ray.

13 Which two of the following can live in both sea water and fresh water?
(i) pike, (ii) salmon, (iii) mackerel, (iv) eel, (v) herring, (vi) roach.
A (i), (iii)   C (iii), (v)
B (ii), (iv)   D (iv), (vi)

14 One of the following is without a cloaca:
A dogfish   C hen
B frog   D rabbit.

15 Which one of the following are viviparous animals?
A frog   C hen
B trout   D rabbit.

16 Which two of the following are filter feeders?
(i) whalebone whale, (ii) penguin, (iii) frog, (iv) herring, (v) toothed whale, (vi) porpoise.
A (i), (iv)   C (iii), (v)
B (ii), (iii)   D (v), (vi)

17 Which two animals in Question 16 are poikilothermic?
A (i), (ii)   C (iii), (iv)
B (ii), (iii)   D (v), (vi)

18 The swim bladder of a fish serves to:
A assist in respiration
B store urinary waste
C alter the body density
D store unfertilized eggs.

19 Which one of the following contains the greater concentration of oxygen for respiration?
A fresh water   C air at sea level
B sea water   D high altitude air.

20 The smallest egg cell amongst the following is produced by a:
A salmon   C human being
B frog   D hen.

21 Which three of the following are respiratory surfaces used for exchange of gases in an *adult* frog?
(i) external gills, (ii) lungs, (iii) internal gills, (iv) skin, (v) intestine, (vi) mouth lining.
A (i), (iii), (v)   C (iii), (iv), (v)
B (ii), (iv), (vi)   D (iv), (v), (vi)

22 Which two of the items in Question 21 are used for exchange of gases in a tadpole whose hind legs are just forming?
A (i), (iii)   C (iv), (v)
B (ii), (iii)   D (v), (vi)

23 Segmentation of the body of a tadpole is clearly seen in the:
A tail muscles   C external gills
B intestine   D eyes.

24 A mature adult male frog can be recognized by its:
A claspers   C green colour
B thumb pads   D penis.

25 One of the following is without eardrums?
A fish   C birds
B amphibia   D mammals.

26 Which of the following is a tetrapod?
A salmon   C newt
B hen   D snake.

27 The main food of adult frogs consists of:
A fruits   C fish
B insects   D leaves.

28 Which three of the following animals show external fertilization?
(i) rabbit, (ii) pigeon, (iii) salmon, (iv) frog, (v) cat, (vi) newt.
A (i), (ii), (v)   C (ii), (v), (vi)
B (i), (iii), (iv)   D (iii), (iv), (vi)

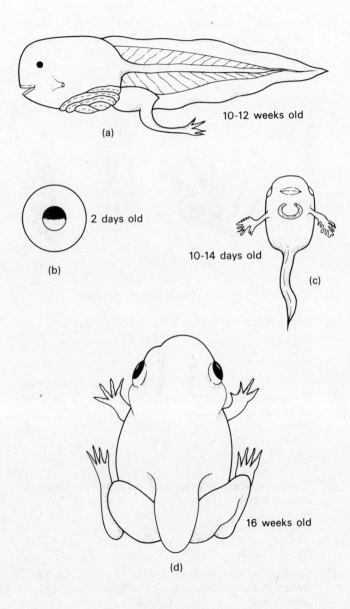

(a) 10-12 weeks old

(b) 2 days old

(c) 10-14 days old

(d) 16 weeks old

Figure 8.1

29 The presence of a long coiled intestine seen in a tadpole with hind legs (Figure 8.1a) indicates that its diet consists mainly of:
A yolk   C animal food
B vegetable food   D animal and vegetable foods.

30 A young 10-day-old tadpole, (Figure 8.1c), which has just emerged from the jelly, clings to the water weed by means of its:
A mouth   C external gills
B operculum   D mucus glands.

31 At the age of three weeks, when the internal gills begin to function, water containing oxygen is drawn into the tadpole's gill chamber by way of the:
A mouth   C spiracle
B nostrils   D sucker.

32 Which one of the following substances will speed up the process of metamorphosis of tadpoles into adult frogs?
A iron   C sodium chloride
B iodine   D calcium.

33 Which one of the following hormone substances also speed up the metamorphosis of frogs?
A adrenaline   C secretin
B thyroxine   D auxin.

34 Three of the following have limbs with webbed digits.
(i) house-fly, (ii) sparrow, (iii) frog, (iv) duck, (v) dog, (vi) bat.
A (i), (ii), (v)   C (ii), (iv), (vi)
B (i), (iii), (v)   D (iii), (iv), (vi)

35 Which one of the following is completely without an exoskeleton?
A cod fish   C frog
B cockroach   D pigeon.

36   The most highly developed sense in a fish is that of:
A smell   C taste
B hearing   D sight.

37   The outer covering of birds consists of:
(i) denticles, (ii) moist skin, (iii) scales, (iv) cuticle,
(v) feathers, (vi) fur.
A (i), (ii)   C (iii), (v)
B (ii), (iv)   D (iv), (vi)

38   Which four of the following terms apply to birds?
(i) poikilothermic, (ii) homoiothermic, (iii) tetrapod,
(iv) feathered, (v) biped, (vi) oviparous, (vii) viviparous,
(viii) furred.
A (i), (iii), (iv), (vii)   C (ii), (iii), (vii), (viii)
B (ii), (iv), (v), (vi)   D (i), (iv), (v), (vii)

39   Which three of the following birds are unable to fly?
(i) hen, (ii) ostrich, (iii) pheasant, (iv) penguin,
(v) canary, (vi) kiwi.
A (i), (iii), (v)   C (iii), (iv), (vi)
B (ii), (iv), (vi)   D (i), (ii), (v)

40   The jaws of birds are covered in:
A denticles   C incisor teeth
B a horny sheath   D many spines.

41   The skin gland of birds secretes one of the following:
A sebum grease   C preen oil
B sweat fluid   D mucus.

42   Which four of the following are contour feathers?
(i) primary wing, (ii) secondary wing, (iii) filoplume,
(iv) down, (v) tail covert, (vi) wing covert.
A (i), (ii), (iii), (iv)   C (iii), (iv), (v), (vi)
B (i), (ii), (v), (vi)   D (ii), (iii), (iv), (v)

43   The young nestling chick is covered entirely by one
type of feather:
A wing feather   C covert feather
B down   D filoplume.

1   2   3   4

Figure 8.2

44   Which feather in Figure 8.2 is called a filoplume?
A 1   C 3
B 2   D 4.

45   How many digits or phalanges are present in the wing
of a bird?
A one   C three
B two   D five.

46   How many digits or phalanges are present in the hind
limb or legs of a sparrow?
A two   C four
B three   D five.

47   The bone of the forelimb or wing of a bird which is
closest to the secondary wing feathers is called the:
A humerus   C ulna
B radius   D phalange.

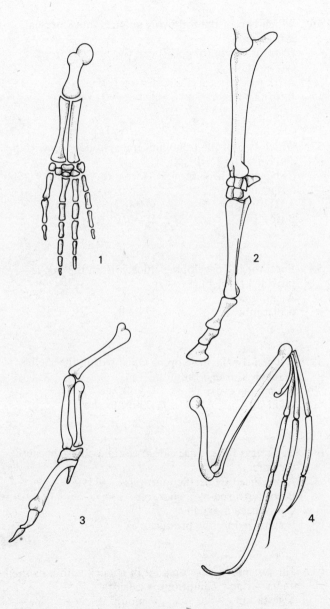

Figure 8.3

48 Which of the pentadactyl limbs shown in Figure 8.3 is that of a bird?
A 1   C 3
B 2   D 4.

49 Which three of the following has a third eyelid or nictitating membrane?
(i) man, (ii) cat, (iii) frog, (iv) fish, (v) bird, (vi) insect.
A (i), (ii), (iv)   C (iii), (iv), (v)
B (ii), (iii), (v)   D (iv), (v), (vi)

50 The breast muscle of a bird, which is responsible for flight movement of the wings, extends from the sternum to one of the following bones:
A coracoid   C humerus
B scapula   D ulna.

51 The number of times one breath of inhaled air passes through the lungs of a bird before being exhaled is:
A once   C continuously
B twice   D three times.

52 Which three of the following give a bird's body a low density?
(i) air sacs, (ii) breast muscle, (iii) streamlined shape, (iv) hollow bones, (v) feathers, (vi) gizzard.
A (i), (iii), (v)   C (ii), (iii), (iv)
B (i), (iv), (v)   D (ii), (iv), (vi)

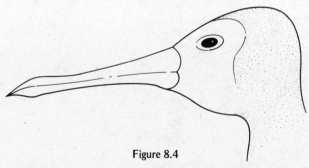

Figure 8.4

53 The beak of the bird shown in Figure 8.4 is suited to feeding upon:
A insects   C fish
B nuts and seeds   D flesh or meat.

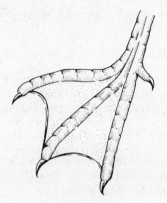

Figure 8.5

54 The foot shown in Figure 8.5 is that of a:
A running bird   C bird of prey
B swimming bird   D game bird.

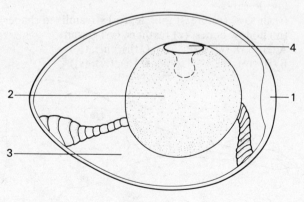

Figure 8.6

55 Which labelled part in Figure 8.6 indicates the position from which the embryo chick develops?
A 1   C 3
B 2   D 4.

56 Which two of the following animals show internal fertilization?
(i) trout, (ii) herring, (iii) cat, (iv) hen, (v) frog, (vi) newt.
A (i), (ii)   C (iv), (v)
B (iii), (iv)   D (v), (vi)

57 Which two of the following embryos develop within a uterus after fertilization?
(i) penguin, (ii) salmon, (iii) frog, (iv) pigeon, (v) rabbit, (vi) man.
A (i), (ii)   C (iv), (v)
B (iii), (iv)   D (v), (vi)

58 Food for the developing chick, within the egg, is provided through the:
A amnion   C yolk sac
B placenta   D shell.

59 Oxygen for the developing chick within the shell is available through the:
A chalaza cords   C yolk sac
B gill slits   D porous shell.

60 Hen's eggs for eating seldom contain chicks because they are:
A produced by parthenogenesis
B not fertilized by a cockerel
C produced asexually
D sterilized by the producer.

61 The process of development of a chick within its shell under suitable conditions is called:
A gestation   C moulting
B incubation   D pupation.

62   The suitable conditions required for the process of
chick development in Question **61** are:
(i) continual temperature of 40 °C, (ii) temperature
of 100° C for one hour, (iii) humid air, (iv) dry air,
(v) 24 hours at a temperature of 0 °C, (vi) exclusion
of air.

A (i), (iii)   C (iv), (v)
B (ii), (vi)   D (iii), (v)

63   How long does it take a domestic fowl chick to develop
from embryo to hatching?

A one day   C two months
B three weeks   D 9 months.

64   Which two of the following animals require some
parental care after birth?
(i) house-fly, (ii) herring spawn, (iii) frog tadpoles,
(iv) thrush chick, (v) domestic fowl chick, (vi) whale
calves.

A (i), (ii)   C (iv), (v)
B (ii), (iii)   D (iv), (vi)

9

Invertebrates

1 Which one of the following is the main means of removing soluble excretory waste from the amoeba?
A food vacuole
C ectoplasmic diffusion
B contractile vacuole
D nucleus.

2 The food vacuole of the amoeba is situated in the:
A endoderm
C endolymph
B endocarp
D endoplasm.

3 Water enters the body of the amoeba from the surrounding fresh water by:
A osmosis
C selective absorption
B diffusion
D transpiration.

4 The osmoregulator of an amoeba is called the:
A food vacuole
C nucleus
B contractile vacuole
D ectoplasm.

5 An amoeba ingests food by:
A sucking
C engulfment
B filtration
D chewing.

6 Oxygen, for the purpose of respiration, enters the amoeba from the surrounding fresh water by:
A diffusion
C food ingestion
B osmosis
D a pseudopodium.

7 An amoeba can be dispersed over considerable distances by means of its:
A cilia
C pseudopodia
B dry cysts
D whip flagella.

8 Which three of the following are cell organelles?
(i) tentacles, (ii) teeth, (iii) mitochondria, (iv) contractile vacuoles, (v) chloroplasts, (vi) pseudopodia.
A (i), (ii), (iii)
C (ii), (iv), (vi)
B (i), (v), (vi)
D (iii), (iv), (v)

9 The process of normal asexual reproduction in the amoeba is called:
A binary fission
C conjugation
B budding
D spore formation.

10 As an amoeba grows larger the surface area and body volume ratio alters to the extent of severely affecting one of the following:
A feeding
C movement
B respiration
D excretion.

11 Which one of the following is a multicellular animal?
A amoeba
C malarial parasite
B hydra
D paramoecium.

12 The stinging cells of the hydra are found in the:
A ectoplasm
C epidermis
B ectoderm
D exoskeleton.

13 Stimuli are conducted through the hydra's body by the:
A cell protoplasm
C axon fibres
B nerve cell net
D rods and cones.

14 Which three of the following animals are hermaphrodite?
(i) amoeba, (ii) hydra, (iii) earthworm, (iv) cabbage white butterfly, (v) tapeworm, (vi) house-fly.
A (i), (iv), (vi)
C (iii), (iv), (v)
B (ii), (iii), (v)
D (iv), (v), (vi)

15　Which three of the following animals show radial symmetry?
(i) earthworm, (ii) butterfly, (iii) sea anemone, (iv) jellyfish, (v) hydra, (vi) snail.
A (i), (ii), (iii)　　　C (ii), (iv), (vi)
B (i), (v), (vi)　　　D (iii), (iv), (v)

16　Bilateral symmetry is seen in animals whose movements are:
A limited or sedentary　C directionless drifters
B aimless floating　　　D in one direction.

17　One of the following animals has a definite circulatory system:
A amoeba　　　C tapeworm
B hydra　　　D earthworm.

18　The oesophagus and intestine is found in one of the following animals:
A hydra　　　C tapeworm
B paramoecium　　　D earthworm.

19　Which one of the following animals is a total parasite?
A hydra　　　C tapeworm
B amoeba　　　D earthworm.

20　Which one of the following animals is without special organs of ingestion?
A hydra　　　C tapeworm
B amoeba　　　D earthworm.

21　Which one of the following is the essential food an earthworm obtains from the soil?
A clay　　　C sand
B plants　　　D lime.

22　Which one of the following is *not* a function of the earthworm in improving the soil?
A aeration　　　C acidification
B drainage　　　D making humus.

23　One of the following will bring living earthworms to the surface of a lawn in the daytime:
A bright sunshine　C rolling or stamping
B heavy rain　　　D raking.

24　An earthworm becomes short and fat by one of the following:
A loss of water through excretory tubes
B contraction of longitudinal muscle
C contraction of circular muscle
D elongation of intestine.

25　Respiration in an earthworm is conducted mainly through one of the following:
A nephridium　　　C epidermis
B clitellum　　　D mouth lining.

26　Which one of the following animals can use a similar method of respiration to the earthworm?
A tapeworm　　　C frog
B house-fly　　　D pigeon.

27　If one living organism preys on another living organism, the organism providing the food is called a:
A parasite　　　C host
B saprophyte　　　D scavenger.

28　Mutualism, or the beneficial relationship between living organisms, is found between:
A algae in the green hydra
B tapeworms in the intestine of pigs
C mushrooms growing in manure
D rats in a refuse heap.

29　Certain microbes are responsible for producing vitamin B in the intestine of man, these are therefore called:
A parasites　　　C symbionts
B saprophytes　　　D scavengers.

Figure 9.1

30 The head end of a tapeworm shown in Figure 9.1 is called a:
A segment   C scolex
B proglottis   D rostellum.

31 A live tapeworm is prevented from being digested and destroyed by the intestinal juices of man by virtue of its:
A own digestive juices   C chitin cuticle
B anti-enzyme secretion   D egg shell gland.

32 Which one of the following foods is most liable to cause tapeworm infection in man?
A fruits   C milk
B meat   D vegetables.

33 The method of respiration used by the tapeworm is:
A aerobically through the skin
B aerobically through tracheoles
C aerobically through its lungs
D anaerobically without oxygen.

34 The fertilized egg of a tapeworm develops into:
A round-worm   C thread-worm
B bladder-worm   D flat-worm.

35 An adult tapeworm attaches itself to the wall of the intestine of man by means of its:
A tentacles   C stylets
B suckers   D proboscis.

36 Which four of the following animals have both jointed limbs and segmented bodies?
(i) earthworm, (ii) spider, (iii) butterfly, (iv) tapeworm, (v) centipede, (vi) crab.
A (i), (ii), (iii), (iv)   C (ii), (iii), (v), (vi)
B (i), (ii), (iv), (vi)   D (iii), (iv), (v), (vi)

37 Which three of the following have bodies divided into a distinct head, thorax and abdomen?
(i) millipede, (ii) housefly, (iii) dogfish, (iv) rabbit, (v) spider, (vi) butterfly.
A (i), (iii), (v)   C (i), (ii), (iv)
B (ii), (iv), (vi)   D (iii), (v), (vi)

38 Which three of the following have three pairs of legs attached to the thorax?
(i) centipede, (ii) lobster, (iii) spider, (iv) cockroach, (v) honey bee, (vi) moth.
A (i), (ii), (iii)   C (iii), (iv), (v)
B (ii), (iii), (iv)   D (iv), (v), (vi)

39 Which three of the following animals show ecdysis?
(i) cockroach, (ii) snake, (iii) crab, (iv) snail, (v) sea
anemone, (vi) hydra.

A (i), (ii), (iii)   C (iii), (iv), (v)
B (ii), (iii), (iv)   D (iv), (v), (vi)

Figure 9.2

Question **40** refers to Figure 9.2 showing the external
features of a locust.

40 What is the part labelled X called?
A head   C thorax
B neck   D abdomen.

41 The animal shown in Figure 9.3 is:
A an insect   C a spider
B a crustacean   D a mollusc.

42 Which two of the following insects have only *one pair*
of wings?
(i) butterfly, (ii) house-fly, (iii) honeybee, (iv) aphid,
(v) mosquito, (vi) moth.
A (i), (iv)   C (iii), (vi)
B (ii), (v)   D (iv), (vi)

Figure 9.3

43 The organs of respiration of an insect have surface
openings called:
A nephridiopores   C spiracles
B stomata   D nares.

44 The insect respiratory system ends in the body tissues
as very fine tubes which are called:
A tracheoles   C capillaries
B nervures   D arterioles.

45 Air is drawn into the respiratory system of an insect by:
A the proboscis   C fan action of wings
B abdomen wall   D movement of the
movement   diaphragm.

46 Leg and wing movements in insects are brought about
by:
A extensor and flexor   C vertical muscles
muscles
B circular muscles   D muscular-epithelial cells.

47 The muscles of insects are attached to points that are:
A outside the   C within the endoskeleton
exoskeleton
B within the exoskeleton D outside the endoskeleton.

48   Due to the possession of circular and longitudinal muscles, one of the following animals can perform different movements to the others.
A amoeba   C house-fly
B earthworm   D hydra.

49   The adult insect formed after complete metamorphosis is called the:
A larva   C nymph
B pupa   D imago.

50   When the egg of an insect hatches directly into a miniature version of the adult, the young insect is called:
A an imago   C a larva
B a pupa   D a nymph.

51   Incomplete metamorphosis is shown in the life cycles of two of the following insects.
(i) butterfly, (ii) aphid, (iii) moth, (iv) mosquito, (v) cockroach, (vi) caddis-fly.
A (i), (iii)   C (iii), (vi)
B (ii), (v)   D (iv), (vi)

52   Which one of the following is an insect larva?
A thread-worm   C ringworm
B wood-worm   D slow-worm.

53   Two of the following are carnivorous feeders?
(i) male mosquito, (ii) female mosquito, (iii) ladybird, (iv) honeybee, (v) ant, (vi) butterfly.
A (i), (ii)   C (i), (v)
B (ii), (iii)   D (iv), (vi)

54   Which three of the following insects have biting jaws for feeding purposes?
(i) house-fly, (ii) cockroach, (iii) aphid, (iv) moth, (v) locust, (vi) butterfly caterpillar.
A (i), (iii), (iv)   C (iii), (iv), (v)
B (ii), (v), (vi)   D (iv), (v), (vi)

55   Which two of the following insects have mouthparts for piercing and sucking?
(i) aphid, (ii) mosquito, (iii) house-fly, (iv) locust, (v) butterfly, (vi) wasp.
A (i), (ii)   C (iv), (v)
B (iii), (iv)   D (v), (vi)

56   One of the following animals is helpful to man:
A termite   C aphid
B ant   D ladybird.

57   The reason for the usefulness of the insect in Question 56, is that it:
A prevents insects attacking wood
B produces nutritious honey
C eats other harmful insects
D produces silk fibre.

58   Which three of the following transmit diseases to man?
(i) locust, (ii) termite, (iii) housefly, (iv) rat fleas, (v) mosquito, (vi) wasp.
A (i), (ii), (iii)   C (iii), (iv), (v)
B (i), (ii), (vi)   D (iv), (v), (vi)

59   The three diseases transmitted by the insects in Question 58 are:
(i) malaria, (ii) food poisoning, (iii) scurvy, (iv) plague, (v) rickets, (vi) lung cancer.
A (i), (ii), (iv)   C (ii), (iii), (v)
B (iii), (v), (vi)   D (iv), (v), (vi)

60   Which one of the following insects would lay its eggs in rotting waste and refuse?
A aphid   C house-fly
B human lice   D wasp.

61   Which one of the following insects would be the cause of summer diarrhoea in humans if it had come into contact with food?
A bee   C ant
B wasp   D house-fly.

62 Which four of the following are characteristics of social insects?
(i) beneficial to man, (ii) show intelligence, (iii) division of labour, (iv) usually produced from one mother, (v) live in colonies, (vi) are all parasites.
A (i), (ii), (v), (vi)   C (i), (ii), (iv), (v)
B (ii), (iii), (iv), (v)   D (iii), (iv), (v), (vi)

63 Which three of the following are social insects?
(i) bed bugs, (ii) aphids, (iii) termites, (iv) ants, (v) honey bees, (vi) head lice.
A (i), (ii), (iii)   C (iii), (iv), (v)
B (i), (ii), (vi)   D (iv), (v), (vi)

64 One of the following is an example of an insect life-cycle showing complete metamorphosis:
A egg – nymph – imago
B egg – caterpillar – imago
C egg – larva – pupa – imago
D egg – chrysalis – pupa – imago.

65 One of the following insects shows incomplete metamorphosis in its life cycle:
A honey bee   C cockroach
B moth   D house-fly.

66 Which two of the following are the pupa stages in the insects life history?
(i) caterpillar, (ii) maggot, (iii) grub, (iv) chrysalis, (v) cocoon, (vi) larva.
A (i), (ii)   C (iv), (v)
B (ii), (iii)   D (v), (vi)

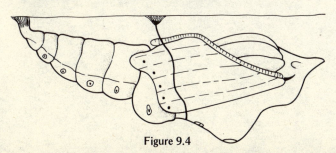

Figure 9.4

67 Figure 9.4 shows a stage in the metamorphosis of a butterfly. This stage is called the:
A imago   C egg
B chrysalis   D larva.

68 Which female insect can produce many daughters without it being fertilized by a male insect?
A butterfly   C moth
B mosquito   D aphid.

69 The type of reproduction in Question **68** is called:
A sexual   C binary fission
B parthenogenesis   D budding.

70 Which one of the following has a pollen basket?
A flower anther   C drone bee
B beehive cell   D worker bee.

71 Prior to taking in food, a house-fly discharges fluid onto the food, the purpose of which is to:
A dissolve and digest the food
B disinfect the food
C sting and kill the food
D test the suitability of the food.

72 Which two of the following animals have segmented bodies?
(i) hydra, (ii) flat-worm, (iii) mosquito, (iv) snail, (v) oyster, (vi) wood-louse.
A (i), (ii)   C (iii), (vi)
B (ii), (iv)   D (iii), (v)

73 Which three of the following have *three* pairs of jointed legs?
(i) slug, (ii) spider, (iii) caterpillar, (iv) centipede, (v) ladybird, (vi) house-fly.
A (i), (ii), (iv)   C (iii), (v), (vi)
B (ii), (iv), (v)   D (iv), (v), (vi)

74 Which two of the following are without wings?
(i) ladybird, (ii) wasp, (iii) spider, (iv) butterfly, (v) mosquito, (vi) wood-louse.
A (i), (v)   C (iii), (vi)
B (ii), (iv)   D (i), (vi)

10

Flowering Plant Morphology

Questions 1—9 refer to the flowering plant in Figure 10.1.

1   This plant takes one year to grow from a seed, to flower, produce fruit, and then die down. Such a plant is called:
    A a woody perennial   C a biennial
    B an herbaceous perennial D an annual.

2   The plant in Figure 10.1 is classified as one of the following because of its form or habit:
    A shrub   C tree
    B herb   D moss.

3   Which one of the following is a flowering plant?
    A moss   C fern
    B toadstool   D buttercup.

4   The reproductive parts of the plant in Figure 10.1 are the:
    A roots   C flowers
    B stem   D leaves.

5   A vegetative part of the plant in Figure 10.1 would be the:
    A fruit   C leaf
    B seed   D petal.

6   Which four of the following are vegetative parts of the flowering plant?
    (i) fruits, (ii) leaves, (iii) roots, (iv) seeds, (v) sepals, (vi) buds, (vii) petals, (viii) stems.
    A (i), (ii), (iii), (iv)   C (ii), (iii), (vi), (viii)
    B (i), (iv), (v), (vii)   D (v), (vi), (vii), (viii)

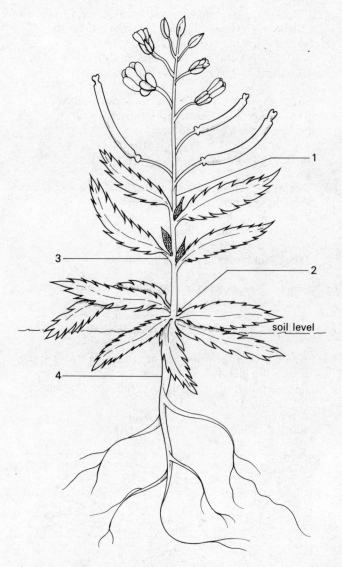

Figure 10.1

7   The stem of the plant in Figure 10.1 is described as being:

A underground   C twining
B erect   D swollen.

8   The point at which the leaf is connected to a stem is called the:

A blade   C axil
B node   D internode.

9   Which labelled part of Figure 10.1 is called the leaf axil?

A 1   C 3
B 2   D 4.

10   Petiole, lamina, midrib, and margin are all component parts of one of the following:

A stem   C leaf
B root   D flower.

11   Select the arrangement of plant parts showing the correct order in which water from the soil reaches the leaf tip.

A midrib – rootlets – stem – taproot – petiole
B petiole – midrib – stem – rootlets – tap root
C rootlets – tap root – stem – petiole – midrib
D tap root – rootlets – stem – midrib – petiole.

12   Which one of the following plants has a swollen tap root?

A meadow grass   C carrot
B oats   D cabbage.

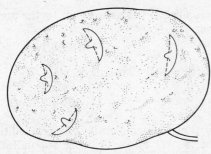

Figure 10.2

13   The reason why a potato tuber (Figure 10.2) is an underground stem, is because it is:

A connected to a stem node
B connected to the root
C has scale leaves with buds
D has many adventitious roots.

14   Which one of the following collective names is given to the male parts of a flower?

A gynaecium   C androecium
B corolla   D calyx.

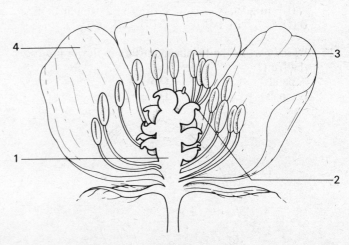

Figure 10.3

15   Where is the receptacle of the flower in Figure 10.3?

A 1   C 3
B 2   D 4.

16   A group or collection of flowers on a branch shoot is called the:

A flora   C floret
B inflorescence   D bouquet.

17 Which one of the following flower parts contain the female gametes?
A filaments   C anthers
B carpels   D sepals.

18 The pollen brought to a flower becomes attached to the:
A style   C carpel
B stigma   D receptacle.

19 Which four of the following plants do not produce or have flowers?
(i) barley, (ii) seaweed, (iii) pine tree, (iv) privet, (v) bracken fern, (vi) mushroom, (vii) holly, (viii) daffodil.
A (i), (iii), (v), (vi)   C (ii), (iv), (vii), (viii)
B (i), (iv), (vii), (viii)   D (ii), (iii), (v), (vi)

20 Which part of the fertilized buttercup flower produces the fruit?
A petals   C ovary
B sepals   D stamens.

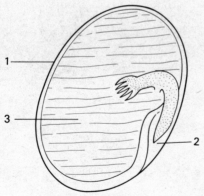

Figure 10.4

21 Questions 21–23 refer to Figure 10.4, which is a sectional view of a pea seed. Which of the following is the name given to part labelled 1?
A pericarp   C testa
B receptacle   D root sheath.

22 Part labelled 3 is called the:
A hypocotyl   C epicotyl
B plumule   D cotyledon.

23 Part labelled 2 is a pore which allows the entry of water into the seed when it is planted, this is called the:
A placenta   C micropyle
B hilum   D testa.

24 Which three of the following plants have two cotyledons in their seeds?
(i) sunflowers, (ii) beans, (iii) cress, (iv) pine trees, (v) grass, (vi) wheat.
A (i), (ii), (iii)   C (iii), (iv), (v)
B (ii), (iii), (iv)   D (iv), (v), (vi)

25 Flowering plants with two cotyledons in their seeds also have one of the following distinguishing features.
A narrow leaves   C fibrous roots
B broad leaves   D parallel leaf veins.

Questions 26–29 refer to Figure 10.5, which is a horse-chestnut twig as it appears in winter.

26 The parts labelled 4 are shield-shaped scars left by the previous years
A flowers   C buds
B leaves   D adventitious roots.

27 The part labelled 1 is the scar left by the previous years:
A flower   C bud
B leaves   D adventitious root.

28 The parts labelled 3 are called:
A lenticels   C leaf scars
B flower scars   D bud scale scars.

29 The parts of the stem, labelled 2 are called the:
A lenticels   C vascular bundles
B stomata   D resin glands.

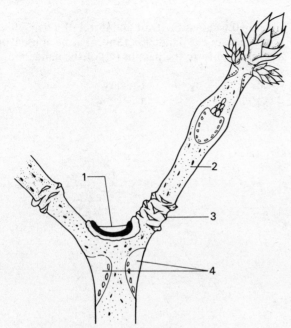

Figure 10.5

30 The bud which is situated between the leaf stalk and the stem is called:

A axillary   C terminal
B intercalary   D internodal.

31 Which three of the following trees are deciduous?
(i) holly, (ii) pine, (iii) sycamore, (iv) larch, (v) privet, (vi) horse-chestnut.
A (i), (ii), (iii)   C (iii), (iv), (v)
B (ii), (v), (vi)   D (iii), (iv), (vi)

32 Which one of the following plants is a biennial?
A turnip   C lettuce
B cabbage   D tomato.

33 Which one of the following plants used in vegetable cookery is a perennial?
A carrot   C cabbage
B mint   D peas.

34 The shoot system of a flowering plant is made up of three of the following:
(i) lateral roots, (ii) root hairs, (iii) tap root, (iv) leaves, (v) stem, (vi) terminal bud.
A (i), (ii), (iii)   C (iii), (iv), (v)
B (ii), (iii), (iv)   D (iv), (v), (vi)

35 The leaves arranged alternately on the plant in Figure 10.1 are of a type named as:
A palmate   C compound
B simple   D pinnate.

36 Which two of the following plant organs or modifications are considered to be 'telescoped' or 'condensed' shoot systems?
(i) iris rhizomes, (ii) daffodil bulbs, (iii) potato tubers, (iv) brussel sprouts, (v) strawberry runners, (vi) crocus corms.
A (i), (vi)   C (iii), (v)
B (ii), (iv)   D (iv), (vi)

37 Which two of the following plant organs, or modifications, are considered to be swollen underground stems.
(i) potato tubers, (ii) strawberry runners, (iii) brussel sprouts, (iv) carrot roots, (v) celery stems, (vi) crocus corms.
A (i), (vi)   C (iii), (v)
B (ii), (iv)   D (v), (vi)

38 Which two of the following plants have flowers consisting of an inflorescence made up of many florets?
(i) daffodil, (ii) tulip, (iii) dandelion, (iv) buttercup, (v) daisy, (vi) sweet pea.
A (i), (ii)   C (iii), (v)
B (ii), (iv)   D (iv), (vi)

39 Which one of the following flowers has a tubular corolla?
A rose   C foxglove
B poppy   D buttercup.

40 Which two of the following plants produce fruits which on drying open by splitting to release their seeds?
(i) apple, (ii) gooseberry, (iii) plum, (iv) broad bean, (v) acorn, (vi) poppy.

A (i), (ii)   C (iii), (v)
B (ii), (iii)   D (iv), (vi)

41 Which two of the following plants produce a one-seeded juicy fruit when ripe?
(i) pear, (ii) tomato, (iii) plum, (iv) strawberry, (v) cherry, (vi) gooseberry.

A (i), (ii)   C (iii), (v)
B (ii), (iv)   D (iv), (vi)

42 Which three of the following fruits are formed from the ovary and also include the *receptacle* which becomes fleshy and juicy?
(i) blackberry, (ii) plums, (iii) strawberries, (iv) apples, (v) pears, (vi) blackcurrants.

A (i), (ii), (iii)   C (ii), (iii), (vi)
B (iii), (iv), (v)   D (iv), (v), (vi)

43 Figure 10.6 shows the fruit and floret of a dandelion. Which part of the dandelion flower has been used, or modified, to form the pappus (X) of the dandelion fruit?

A calyx   C ovary
B corolla   D stigma.

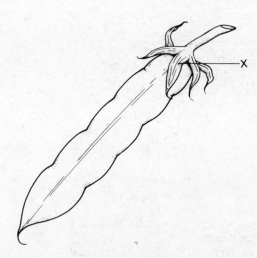

Figure 10.7

44 Which part of the pea flower produces the remains labelled X in Figure 10.7 of the pea pod?

A stigma   C stamens
B sepals   D petals.

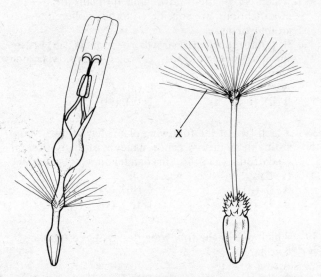

Figure 10.6

45 Which two of the following plants are modified buds which are eaten as vegetables?
(i) cauliflower, (ii) brussel sprouts, (iii) cabbage, (iv) onions, (v) leeks, (vi) lettuce.

A (i), (iii)   C (iii), (v)
B (ii), (iv)   D (iv), (vi)

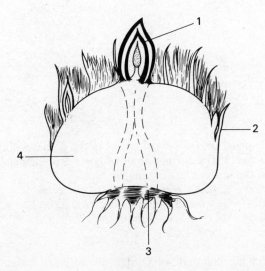

Figure 10.8

50 How many sepals has a buttercup?
  A none   C 5
  B 10   D 3.

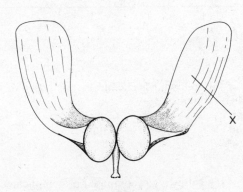

Figure 10.9

Figure 10.9 is a diagram of a sycamore fruit.

46 Figure 10.8 is a transverse view of the crocus corm. Which labelled part will become the future foliage leaves?
  A 2   C 4
  B 3   D 1.

47 Which one of the following terms describes the buttercup flower?
  A sterile   C male
  B hermaphrodite   D female.

48 Which one of the following flowers have petals named as the keel, standard, and wings?
  A buttercup   C dandelion
  B sweet pea   D meadow grass.

49 Which three of the following features are parts of the seed?
  (i) pericarp, (ii) stigma, (iii) testa, (iv) embryo, (v) hilum, (vi) style.
  A (i), (ii), (iii)   C (iii), (iv), (v)
  B (ii), (iii), (iv)   D (iv), (v), (vi)

51 What method is used to disperse the fruit?
  A water   C animals
  B wind   D explosion.

52 What part of the plant in Figure 10.9 is used to form the wing X?
  A testa   C plumule
  B pericarp   D calyx.

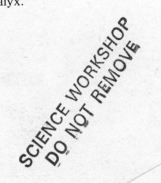

11

Soil

1 One of the following is chemically composed of silica or silicon dioxide.
   A water   C chalk
   B sand   D humus.

2 One of the following is composed of complicated salts called alumino-silicates.
   A chalk   C clay
   B sand   D humus.

Questions 3–5 refer to Figure 11.1.

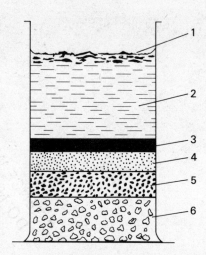

Figure 11.1

3 Soil stirred into water contained in the tall jar will settle into layers. The correct order of layers, commencing with the bottom layer, being:
   A sand – gravel – clay – silt
   B silt – clay – sand – gravel
   C silt – clay – gravel – sand
   D gravel – sand – silt – clay.

4 Which labelled part will contain the soluble food or raw materials of a green plant?
   A 3   C 1
   B 4   D 2.

5 Which labelled part will consist mainly of the dead remains of green plants?
   A 3   C 1
   B 6   D 5.

Questions 6–15 refer to Table 11.1 which shows the analysis of four soils W, X, Y, and Z; each component being given as a percentage by weight.

TABLE 11.1

Soil component	Soil samples			
	W	X	Y	Z
Silica	80	48	65	35
Alumino-silicate	15	14	23	50
Humus	4	30	10	10
Water	1	8	2	5
pH	7	6	7	7

6   Which one of the four soil samples is a very sandy soil?
A. W   C. Y
B. X   D. Z.

7   Which soil sample is a heavy clay soil?
A. W   C. Y
B. X   D. Z.

8   Which soil sample could have been obtained from the depths of a beech or oakwood?
A. W   C. Y
B. X   D. Z.

9   Which soil sample will be sticky and cling to a gardeners boots as clods?
A. W   C. Y
B. X   D. Z.

10   The soil that will drain and dry out the quickest is:
A. W   C. Y
B. X   D. Z.

11   Which one of the following gardening materials is rich in humus?
A lawn sand   C peat
B sulphate of ammonia   D lime.

12   Which soil sample will benefit the most by treating with lime?
A. W   C. Y
B. X   D. Z.

13   One hundred grammes of soil sample Z was heated in a drying oven at 90 °C for two days. What will be the loss in weight of the sample Z?
A 5 g   C 17 g
B 10 g   D 35 g.

14   Another fresh 100 g sample of soil Z was taken and heated for over three hours in a crucible furnace at 1000 °C. What will be the total weight loss of the sample?
A 5 g   C 15 g
B 45 g   D 50 g.

15   The loss in weight of the sample Z heated in a crucible furnace at 1000 °C is due to vapourization of:
(i) sand, (ii) silt, (iii) clay, (iv) humus, (v) gravel, (vi) water.
A (i), (ii)   C (iii), (v)
B (iii), (iv)   D (iv), (vi)

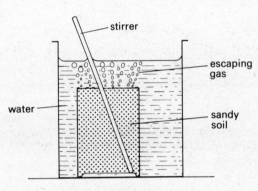

Figure 11.2

16   An open can of fresh sandy soil is placed in the jar of water and is stirred as shown in Figure 11.2, what gas will escape from the soil and rise to the surface?
A carbon dioxide   C air
B ammonia   D marsh gas.

17   In which of the following columns of dry soils would the greatest water rise be seen?
A sandy soil   C peaty loam
B sandy loam   D clayey soil.

18   Which one of the following methods of soil cultivation will help to reduce water loss from the soil?
A rolling   C ploughing
B hoeing   D sieving.

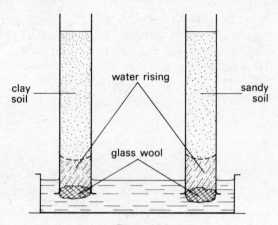

Figure 11.3

19   The upward movement of water seen in the glass tubes of dry soil shown in Figure 11.3 is due to:
A gravity
B capillarity
C osmosis
D diffusion.

20   The effect of adding lime to a soil is to:
A clump the sand grains together
B clump the clay particles together
C feed the plant with calcium
D release ammonia from manures.

21   Which one of the following would decrease the amount of air in the soil?
A heavy flooding
B frequent forking
C earthworms
D moles.

22   Which three of the following require and use air in the soil?
(i) moulds, (ii) moles, (iii) anaerobic bacteria, (iv) aerobic bacteria, (v) rabbits, (vi) root hairs.
A (i), (iv), (vi)
B (ii), (iii), (v)
C (iii), (iv), (v)
D (iv), (v), (vi)

23   What type of soil contains lime or limestone?
A acid
B peaty
C salty
D alkaline.

24   Acid soils form when the soil contains mostly:
A sand
B peat
C salt
D chalk.

25   One of the following can hold water and also provide a nitrogenous food as raw material for green plants and microbes.
A sand
B humus
C clay
D lime.

26   Which three of the following are removed when the soil is heated at 1000 °C for three hours?
(i) phosphates, (ii) humus, (iii) potassium, (iv) microbes, (v) water, (vi) iron.
A (i), (ii), (vi)
B (ii), (iv), (v)
C (iii), (iv), (vi)
D (iv), (v), (vi)

27   Which one of the following require sterilized soil?
A compost heaps
B heated greenhouses
C garden lawns
D vegetable gardens.

Questions 28–30 refer to Figure 11.4

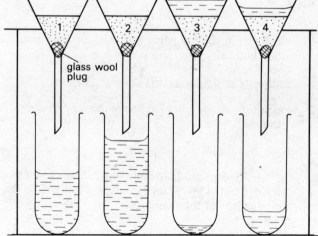

Figure 11.4

28   Equal volumes of water are poured onto equal volumes of different fresh soil samples supported in the filter funnels plugged with glass wool. Which funnel will contain the sandiest soil?
A 1
B 2
C 3
D 4.

29 The passage of the water through the soil in the funnels demonstrates one of the following:
A diffusion   C capillarity
B permeability   D osmosis.

30 The water passes through the soil to reach the test-tube beneath by the effect of:
A surface tension   C gravity
B hygroscopy   D air pressure.

31 Which one of the following samples of soil will produce the greatest growth of microbes when added to sterile agar nutrient cultures, and allowed to incubate in a warm place for 7 days?
A garden loam treated with high pressure steam
B dry sandy soil treated with pig manure
C baked sandy soil treated with sulphate of potash
D garden loam soil watered with formaldehyde solution.

32 The ineffective or incomplete washing of soil-contaminated salad vegetables (such as lettuce), may cause infection of humans with parasites of:
A lock-jaw disease   C tapeworms
B round-worm   D liver fluke.

33 Which one of the following soils will warm up at a more rapid rate than the others in the early spring sunshine?
A yellow clay soil   C black peaty soil
B brown sandy soil   D red loam.

34 If a small amount of hydrochloric acid is poured onto a sample of soil in a watch glass, and bubbles of gas, or foam, are produced. This test indicates that the soil:
A contains escaping air   C requires liming
B contains limestone   D is short of humus.

35 One of the following soil top dressings will help to absorb heat into the soil when applied to its surface.
A silver sand   C chimney soot
B sawdust   D slaked lime.

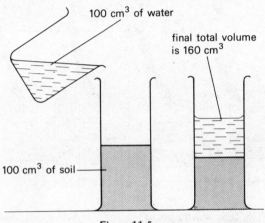

100 cm³ of water

final total volume is 160 cm³

100 cm³ of soil

Figure 11.5

36 One hundred cubic centimetres of water are added to 100 cm³ of soil, after stirring the soil and water together the final total volume is 160 cm³. (Shown in Figure 11.5.) What is the volume of air contained in the soil?
A 200 cm³   C 40 cm³
B 160 cm³   D 80 cm³.

37 Some cabbage leaves are dried then heated strongly in a crucible at a temperature of 1000 °C until a white ash remains. Which two of the following would be components of the ash?
(i) glucose, (ii) potassium, (iii) amino acids, (iv) carbon, (v) phosphorus, (vi) ammonia.
A (i), (iv)   C (iii), (vi)
B (ii), (v)   D (iv), (vi)

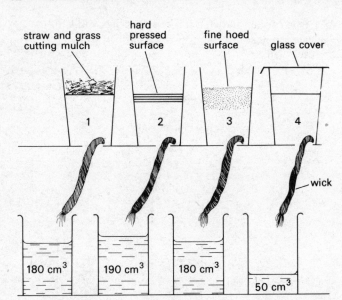

straw and grass cutting mulch

hard pressed surface

fine hoed surface

glass cover

1   2   3   4

wick

180 cm^3 190 cm^3 180 cm^3

50 cm^3

Figure 11.6

38   The four plastic pots (shown in Figure 11.6) all contain a moist, peaty loam soil with a wick dipping into a beaker of water containing originally 200 cm^3 water. Each pot of peaty loam is subjected to different treatment as indicated. After 7 days the beakers were removed from the pots. To which pot was the beaker containing 50 cm^3 of water, connected?
A 1   C 3
B 2   D 4.

39   If water culture solutions are used to grow seedlings it will be necessary to:
A change the solution daily
B blow air through the solution
C remove the jar black-out cover daily
D add disinfectant to the solution.

40   Which of the following is one of the main elements needed for healthy dark-green leaves and the prevention of *chlorosis*?
A aluminium   C iron
B sodium   D tin.

41   Another major element responsible for healthy dark-green leaves and the prevention of chlorosis is:
A chlorine   C magnesium
B mercury   D zinc.

42   Which one of the following are groups of major elements, (also called macronutrients) and required for healthy normal growth of flowering plants.
A nitrogen, phosphorus, potassium, lead, mercury, arsenic, iron
B calcium, potassium, magnesium, iron, chlorine, bromine, nitrogen
C calcium, sulphur, magnesium, nitrogen, iron, phosphorus, potassium
D nitrogen, iron, phosphorus, calcium, iodine, lead, mercury.

43   In addition to the major elements (or macronutrients), others called trace elements (or micronutrients) are needed by healthy plants. Which of the following is a trace element?
A nitrogen   C potassium
B phosphorus   D copper.

44   Which one of the following soil manures and fertilizers provides a plant with nitrogen, phosphorus, and potassium?
A sulphate of ammonia   C farmyard manure
B superphosphate   D soot.

45   Which one of the following provides a soil with a readily available supply of soluble nitrogen?
A nitrate of soda   C hoof and horn
B dried blood   D fish meal.

46   Which one of the following is harmful to a plant or soil when applied in large amounts?
A bone meal   C sulphate of ammonia
B seaweed   D garden compost.

47 Which two of the following crops can increase the nitrogen content of a soil?
A cabbages and broccoli   C oats and barley
B peas and beans   D turnips and beets.

48 Ploughing in the plant roots improves the nitrogen content of a soil and is part of the process of:
A crop rotation   C food chains
B sillage making   D intensive cultivation.

49 Sulphate of ammonia, when added to a soil, must be changed to a more suitable form before absorption by a green plant. This is achieved in the soil by:
A denitrifying bacteria   C nitrifying bacteria
B nitrogen-fixing   D moulds.
   bacteria

50 The nitrogen of a dead plant or animal is locked-up in one of the following body components:
A fats   C proteins
B minerals   D carbohydrates.

51 Which one of the following is added to the soil during a thunderstorm by lightning flashes?
A nitrates   C iron
B sulphates   D calcium.

52 Certain plants have tiny swellings on their roots called root nodules. Which of the following plants are affected in this way?
A turnips   C brussel sprouts
B clover   D lettuces.

53 The root nodules described in Question 52 are inhabited by one of the following:
A fungi   C algae
B bacteria   D protozoa.

54 The inhabitants of root nodules are examples of a relationship between living organisms called:
A parasitism   C mutualism
B commensalism   D saprophytism.

55 The inhabitants of root nodules manufacture one of the following materials:
A fats   C carbohydrates
B proteins   D vitamins.

56 One of the raw materials used by the root nodule inhabitants in their manufacturing process is drawn from the soil, namely:
A nitrites   C nitrogen
B ammonia   D nitrates.

57 Plant and animal remains will decay more rapidly in one of the following types of soil.
A sandy   C waterlogged
B acidified   D heavily rolled over.

58 The brownish-black, spongy solid formed as the end product of decay of plant and animal remains is called:
A manure   C silt
B humus   D coal.

59 The main form in which the majority of green flowering plants receives their nitrogen is:
A from humus through the roots
B as nitrogen gas through the leaves
C as soil nitrates through the roots
D by the root nodule association.

60 Insectivorous plants grow in soil deficient in nitrogen. Their supplies of essential nitrogen come from:
A denitrifying bacteria   C thunderstorms
B root nodules   D insect food.

61 One of the following is an insectivorous plant:
A buttercup   C rose
B butterwort   D apple.

62 Insects are caught and trapped by the plants in Question 61 by means of their:
A roots   C tentacles
B leaves   D flowers.

Plant Structure and the Importance of Water

1   Which one of the following would have a water content
    of about 90 per cent by weight?
    A hazel nuts   C lettuce leaves
    B dried peas   D raisins.

2   Which two of the following fruits contain about 90 per
    cent water by weight?
    (i) tomatoes, (ii) chestnuts, (iii) walnuts, (iv) prunes,
    (v) strawberries, (vi) raisins.
    A (i), (iii)   C (ii), (iv)
    B (i), (v)   D (ii), (vi)

3   What is the average water content of dried fruits, dried
    seeds, and dried vegetables?
    A 0–20%   C 40–60%
    B 20–40%   D 60–80%.

4   The part of the plant having the lowest water content is
    the:
    A stem   C leaf
    B root   D seed.

Questions 5–12 refer to Figure 12.1 which represents a cell
of onion scale leaves, surrounded by other cells as seen
through a microscope.

5   If the cell in Figure 12.1 were separated from the other
    surrounding cells its shape would appear to be:
    A flattened and thin like a biscuit,
    B swollen and distended like a balloon
    C box-shaped with sharp corners
    D shapeless.

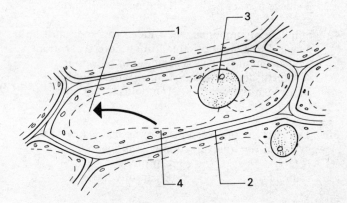

Figure 12.1

6   The wall of the cell part labelled 2, is made of:
    A cutin   C cellulose
    B mucilage   D chitin.

7   Which labelled part of the onion plant cell is non-living
    material?
    A 4   C 2
    B 1   D 3.

8   The element calcium is found mainly within the:
    A cell wall   C nucleus
    B middle lamella   D cytoplasm.

9   Which labelled part contains chloroplasts, mitochondria,
    and starch grains?
    A 4   C 2
    B 1   D 3.

10 If the cell structure in Figure 12.1 was also typical of red beetroot cells, which labelled part would contain the red pigment?
A 4   C 2
B 1   D 3.

11 Cells taken from onion, daffodil, and tulip bulbs are similar to Figure 12.1. In which labelled part would the main and real differences between these different plants be found?
A 1   C 3
B 2   D 4.

12 Which three of the following are components of part labelled 1?
(i) starch, (ii) protein, (iii) cellulose, (iv) chlorophyll, (v) glucose, (vi) water, (vii) potassium, (viii) fat.
A (i), (ii), (vi)   C (iv), (v), (vii)
B (iii), (iv), (viii)   D (v), (vi), (vii)

13 Which one of the following have mucilage in their cell walls?
A fungi   C mosses
B algae   D ferns.

14 The mucor mould plant body does not show a division into cells compared to spirogyra. The mucor plant body is:
A multicellular   C coenocytic
B unicellular   D non-nucleate.

15 From what part of the orange fruit does orange juice come from:
A nucleus   C vacuole
B cytoplasm   D middle lamella.

16 Pectin (essential for the setting of jams and jellies) comes from the part of the plant cell called the:
A nucleus   C vacuole
B cytoplasm   D middle lamella.

17 The labelled part of the cell shown in Figure 12.1 that provides roughage in a human's diet is:
A 4   C 2
B 1   D 3.

18 A high concentration of soluble glucose is found close to part 4 and moves to point 1 in Figure 12.1, where there is a low concentration of glucose. The direction of movement is shown by the arrow. What is this movement of glucose within the cell sap called?
A translocation   C convection
B osmosis   D diffusion.

19 The wall of a plant cell allows water, sugars, and salts to pass through it both inwardly and outwardly. Such a cell wall is said to be:
A permeable   C non-porous
B semi-permeable   D impervious.

20 The movement of the water, sugars, and salts through the cell wall of a plant from the cell sap outwards to its surroundings is called:
A translocation   C diffusion
B radiation   D convection.

21 The type of movement of solutes and water from the cell sap and out through the cell wall is seen for example:
A in living spirogyra in pond water
B when strawberries are sprinkled with sugar
C in living yeasts in glucose solution
D when raisins swell up in water.

22 An example of diffusion amongst the following is:
A water rising from the roots to the leaves of plants
B sugars produced in shoots being moved to roots for storage
C carbon dioxide entering a plant from the air
D water passing from warm soil into the air.

Questions **23–31** refer to Figure 12.2.

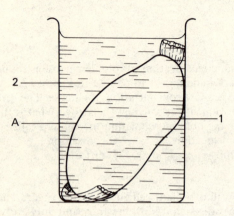

Figure 12.2

23   A transparent bag full of the solution 1 is placed in a beaker of liquid 2 and left for two days; after which period the bag became swollen and had increased in weight. What is the liquid 2?
A water   C ether
B ethanol (ethyl alcohol) D acetone.

24   The bag containing the solution is made of a material A, which is:
A glass   C cellulose
B rubber   D perspex.

25   The movement of liquid 2 into solution 1 through the membrane A is called:
A diffusion   C capillarity
B osmosis   D absorption.

26   A portion of the solution 1 when heated with Fehling's solution, produces a reddish-green precipitate. This indicates that the solution 1 is a solution of:
A sucrose and ethyl   C sucrose and paraffin
   alcohol
B glucose and ether   D glucose and water.

27   The liquid 2 is non-inflammable and is the solvent for solution 1. The solvent for solution 1 is therefore:
A ethyl alcohol   C acetone
B ether   D water.

28   When liquid 2 is tested with Fehling's solution, after the bag of solution 1 has been immersed in it for 2 days, a reddish green precipitate was obtained. The reason for this is:
A osmosis of the solute into 2
B diffusion of the solute into 2
C absorption of the solute into 2
D capillary attraction of solute into 2.

29   In another experiment (using the apparatus shown in Figure 12.2) a solution of 20 per cent glucose was placed in the bag and 10 per cent glucose solution placed in the beaker. What will be the observed result of leaving the bag of solution in the solution in the beaker for two days? The solution in the bag will:
A swell out   C show no change
B shrivel up completely D show slight shrinkage.

30   If a bag containing a solution, consisting of 5 per cent sucrose solution, was placed in a beaker containing a solution consisting of 20 per cent glucose, the observed result would be that the solution in the bag will:
A swell out   C show no change
B shrivel up   D burst.

31   The special membrane A used in this experiment, which allows the passage of small molecules but not larger molecules, is described as being:
A permeable   C porous
B selectively permeable D honeycombed.

32   Four pieces of freshly cut potato (of equal size and shape) were placed in the liquids shown in Figure 12.3. Which liquid will make the potato feel hard and firm in contrast to being soft and flabby after immersion for 24 hours?
A 1   C 3
B 2   D 4.

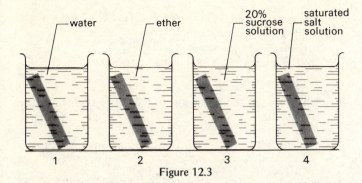

Figure 12.3

33 Four pieces of well-washed, fresh red beetroot (of equal size and shape), are placed in the liquids shown in Figure 12.3. Which beaker of liquid will develop a red colour after 30 minutes?

A 1   C 3
B 2   D 4.

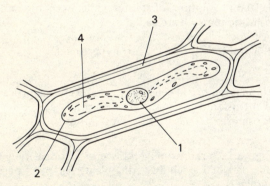

Figure 12.4

34 Figure 12.4 is an onion leaf cell immersed in strong sucrose solution, and the cell contents are shrunk and detached from the cell wall. Which labelled part plays the direct role as a selectively permeable membrane during osmosis?

A 1   C 2
B 3   D 4.

35 The onion cell in Figure 12.4, after immersion in the sugar solution, is in a state of:

A dehydration   C plasmolysis
B turgor   D diffusion.

36 Which one of the following fluids, when watered onto the leaves of a green flowering plant, will cause them to wilt?

A tap water   C seaweed manure solution
B salt water   D distilled water.

37 The cell membrane of an amoeba is:

A impermeable   C permeable
B selectively permeable D porous.

38 The organ mainly concerned with osmosis within the amoeba cell is the:

A nucleus   C contractile vacuole
B food vacuole D pseudopodium.

39 A spirogyra cell takes in essential salts and water by means of its:

A chloroplast   C cytoplasm
B pyrenoid   D nucleus.

40 Which one of the following can take in water by osmosis?

A dried peas   C dried raisins
B dry biscuits   D dry wool and hair.

41 In order that bottled or canned fruits remain firm and intact and do not shred or burst, it is essential to:

A boil them well before bottling and canning
B use over-ripe fruits
C use the correct sugar syrup
D use fresh water only.

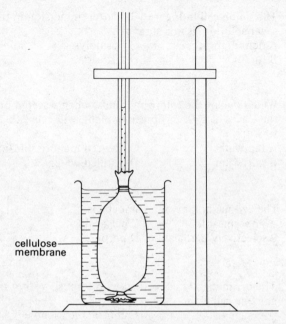

cellulose membrane

Figure 12.5

42 The apparatus shown in Figure 12.5 is to demonstrate:
A root pressure   C transpiration pressure
B osmotic pressure   D atmospheric pressure.

43 Which one of the following is the best definition of osmosis:
A movement of molecules from regions of high concentration to regions of low concentration
B movement of a solvent across a selectively permeable membrane
C movement of solution upwards through fine hair-like tubes
D movement of plant organs in response to an external directional stimulus.

44 Which three of the following are selectively permeable?
(i) aluminium foil, (ii) polythene plastic, (iii) small

intestine lining, (iv) lung alveolus, (v) leaf cuticle, (vi) kidney tubules.
A (i), (ii), (iii)   C (iii), (iv), (vi)
B (ii), (iv), (v)   D (iv), (v), (vi)

45 Which one of the following is an osmoregulator organ in the human body?
A pancreas   C liver
B kidneys   D salivary gland.

46 Which three of the following are processes of diffusion?
(i) salting of herring, (ii) salt excretion in kidney, (iii) sugaring candied peel, (iv) oxygen absorption in lungs, (v) glucose absorption in intestine, (vi) crisping lettuce in water.
A (i), (ii), (iii)   C (ii), (iii), (vi)
B (ii), (iv), (v)   D (iv), (v), (vi)

47 Different solutions of solutes which exert the same osmotic pressure are called:
A tonic   C hypotonic
B isotonic   D hypertonic.

48 Human red blood cells take in water by the process of:
A diffusion   C adsorption
B osmosis   D imbibition.

49 Figure 12.6 represents a transverse sectional view of one part of a green flowering plant. The plant part shown is a:
A root   C leaf
B stem   D bud.

50 The plant part in Question **49** is recognized because of:
A the presence of many root hairs
B vascular bundles arranged circularly
C the central xylem
D many starch grains in the pith.

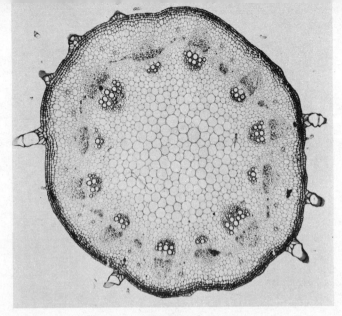

Figure 12.6

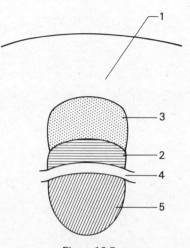

Figure 12.7

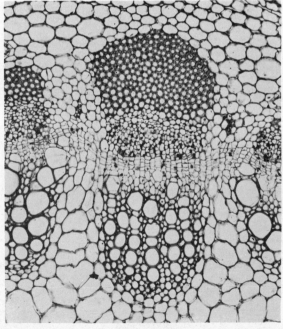

Figure 12.8

Questions **51–54** refer to Figures 12.7 and 12.8 which show details of a vascular bundle featured in Question **49**.

51 Which labelled part of the vascular bundle will conduct water and salts to the leaf?

A 2   C 5

B 3   D 4.

52 The phloem or bast is part labelled:

A 2   C 5

B 3   D 4.

53 The strengthened cells providing mechanical support for the plant are:

A 3 and 4   C 2 and 4

B 2 and 5   D 3 and 5.

54 The part labelled 1 is called the:

A medulla   C cortex

B pith   D epidermis.

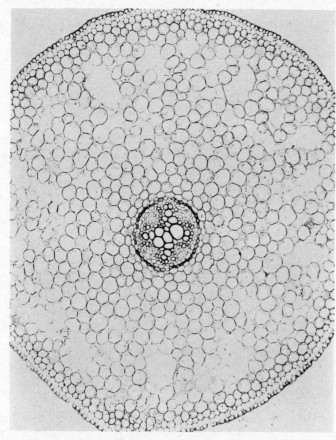

Figure 12.9

55 Figure 12.9 is a transverse sectional view of part of a green flowering plant. The plant part shown is a:

A root   C leaf

B stem   D bud.

56 The specimen shown in Figure 12.9 is recognized because of:

A the root hairs

B central vascular tissue

C scattered vascular bundles

D green chloroplasts in the pith.

57 The structural feature of the specimen in Figure 12.9 is well constructed to withstand a force of:

A bending   C tearing

B pulling   D compression.

Questions **58–60** refer to Figure 12.10 and 12.11 which are parts of a green flowering plant seen in a transverse sectional view.

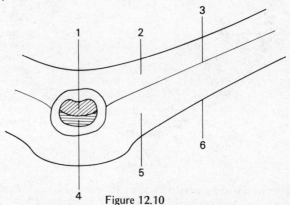

Figure 12.10

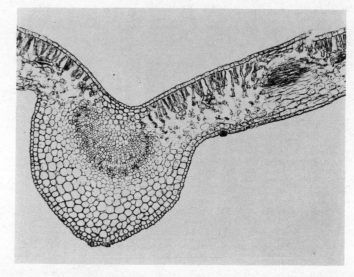

Figure 12.11

58   Where are the palisade cells found?
      A 1   C 5
      B 2   D 6.

59   The cuticle layer is produced by the:
      A epidermis   C phloem
      B xylem   D palisade.

60   Where are the majority of stomata found in the specimen?
      A 3   C 2
      B 6   D 5.

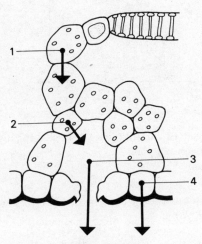

Figure 12.12

61   Parts labelled 1, 2, 3, and 4 in Figure 12.12 indicate the possible paths of water as it passes through the leaf mesophyll from the vein to the exterior air. Which labelled part shows movement of water by osmosis?
      A 4   C 2
      B 1   D 3.

62   Which labelled part in Figure 12.12 shows movement of water by diffusion?
      A 4   C 2
      B 1   D 3.

63   Which one of the following parts of a root provide the greatest individual total surface area for absorption purposes?
      A main taproot   C root hairs
      B rootlets   D branch roots.

64   Which three of the following can enter the root hairs of a plant from the soil?
      (i) proteins, (ii) carbon dioxide, (iii) starch, (iv) oxygen, (v) humus, (vi) water.
      A (i), (iii), (v)   C (iii), (iv), (v)
      B (ii), (iv), (vi)   D (i), (ii), (vi)

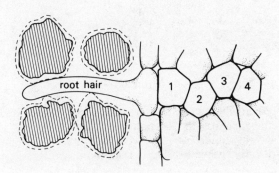

Figure 12.13

65   What is considered to be the main method by which water travels from the root hair into the centre of the root, by cells labelled 1, 2, 3, and 4 shown in Figure 12.13?
      A diffusion   C osmosis
      B evaporation   D filtration.

66   The apparatus shown in Figure 12.14, consisting of a manometer gauge connected to a cut stem of a vine, is used to measure:
      A osmotic pressure   C root pressure
      B transpiration rate   D growth rate.

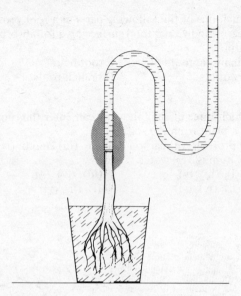

Figure 12.14

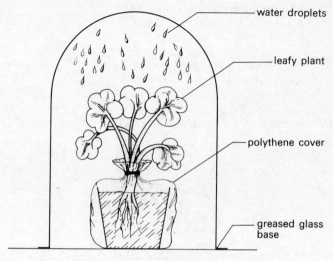

Figure 12.15

67 When will the greatest rate or pressure be recorded by the apparatus in Question **66**?
A mid-winter   C late summer
B early spring   D autumn.

68 Which one of the following chemicals could be used to show that the droplets of liquid condensed inside the bell jar, in Figure 12.15, are of water?
A lime water   C anhydrous calcium chloride
B litmus solution   D anhydrous copper sulphate

69 What would be the colour change observed when using the correct chemical reagent for the test in Question **68**?
A to red   C to white
B to blue   D none.

70 Which one of the following would be unable to release the water shown in Figure 12.15?
A soil   C leaves
B flowers   D stem.

71 Which one of the following is a place for loss of water from the green plant?
A petiole   C epidermis
B midrib   D vein.

72 A deciduous tree in winter loses water by way of one of the following:
A bark   C buds
B leaves   D flowers.

73 When cobalt chloride paper becomes damp, it will show a colour change as follows:
A white to blue   C yellow to pink
B red to blue   D blue to pink.

74 Small squares of cobalt chloride paper are fixed with transparent sticky tape to the following parts of a potted flowering plant. (The sticky tape excludes the air from the cobalt chloride paper.) Which part of the plant will be the first to change the colour of the cobalt chloride paper?
A upper leaf surface   C stem
B lower leaf surface   D petal.

75 Which of the following leaf cells are without chloro-
plasts?
A palisade cells   C guard cells
B mesophyll cells   D epidermal cells.

76 Which one of the following plants has the greatest
number of leaf pores on its upper surface?
A horse-chestnut   C daffodil
B water-lily   D beech.

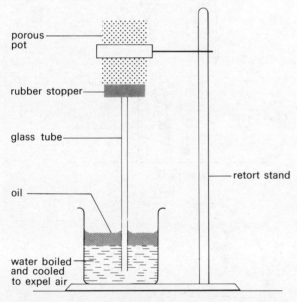

Figure 12.16

77 At which part of the apparatus (an atmometer) shown
in Figure 12.16 can water loss occur?
A through the oil layer   C the porous pot
B the glass tube   D the rubber stopper.

78 The water loss in the apparatus in Figure 12.16 will
take place by the physical process of:
A evaporation   C sublimation
B condensation   D boiling.

79 The loss of water from the porous pot can be
compared to the process of water loss by one of the
following leaf cells:
A spongy mesophyll   C stomata
B phloem vessel   D palisade.

80 Which part of the apparatus in Figure 12.16
corresponds to the xylem vessel of a plant stem
and root?
A beaker of water   C porous pot
B glass tube   D retort stand.

81 The process of loss of water from the surface of a
plant is called:
A translocation   C respiration
B transpiration   D photosynthesis.

82 Which one of the following processes will produce
water in a green flowering plant?
A photosynthesis   C respiration
B translocation   D excretion.

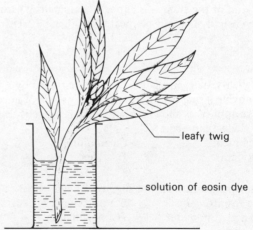

Figure 12.17

83 What special precautions were used in obtaining the leafy
twig shown in Figure 12.17? It was necessary to cut it:
A at night   C under water
B in bright sunshine   D after watering the bush.

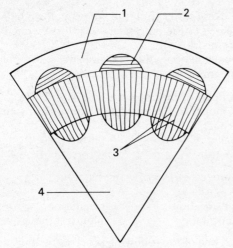

Figure 12.18

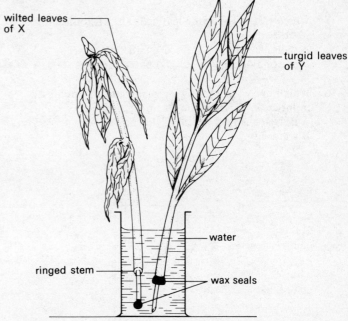

Figure 12.19

84   After leaving the leafy twig of Question **83** in the
     eosin dye solution in a warm room for eight hours, a
     section of the stem was cut and examined. Where would
     the eosin dye colour be located in Figure 12.18?
     A 4   C 1
     B 3   D 2.

85   The leafy twigs in Figure 12.19 were ringed by
     removing a ring of the stem's outer tissues. Wax seals
     were applied as shown. What vascular tissue is sealed
     by the wax in plant Y?
     A phloem   C cambium
     B xylem   D cortex.

86   The apparatus shown in Figure 12.20 is called:
     A a respirometer C a potometer
     B an auxonometer D a porometer.

87   The apparatus in Figure 12.20 measures one of the
     following *directly*:
     A rate of transpiration   C rate of respiration
     B rate of water   D rate of growth.
       absorption

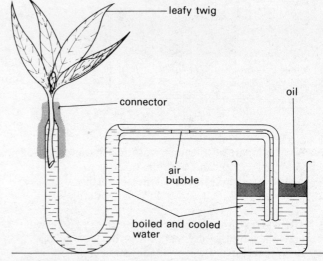

Figure 12.20

88 *Indirectly* the apparatus in Figure 12.20 can show the affect of various external effects on:
A rate of transpiration   C rate of respiration
B rate of water   D rate of growth.
  absorption

89 Select the correct order showing the path of water from roots to the leaves in a green plant.
A root hair – phloem – cortex – mesophyll – epidermis
B cortex – root hair – phloem – palisade – epidermis
C root hair – cortex – xylem – mesophyll – stomata
D epidermis – cortex – xylem – palisade – stomata.

90 Select the correct order of physical processes by which water passes through a green plant from soil water to air.
A capillarity – osmosis – diffusion – evaporation
B osmosis – capillarity – evaporation – diffusion
C diffusion – evaporation – capillarity – osmosis
D evaporation – diffusion – osmosis – capillarity.

91 Leafy shoots fixed in the apparatus shown in Figure 12.20 are placed in greenhouses providing the following external conditions. In which greenhouse will the fastest rate of air bubble movement be seen in the apparatus.
A warm, humid, poorly ventilated
B warm, dry, well ventilated
C cold, damp, poorly ventilated
D cold, blackened out, well ventilated.

92 Which three of the following leaf modifications are ways of reducing water loss in green plants?
(i) large broad leaves, (ii) thin cuticles, (iii) thick waxy cuticles, (iv) compound leaves, (v) hairy leaves, (vi) needle-shaped leaves.
A (i), (ii), (iii)   C (iii), (v), (vi)
B (ii), (iii), (iv)   D (i), (iv), (vi)

93 Which one of the following plants have needle-shaped leaves?
(i) buttercup, (ii) dock, (iii) gorse, (iv) heather, (v) beech, (vi) larch.
A (i), (ii), (iv)   C (iii), (iv), (vi)
B (ii), (iii), (v)   D (i), (v), (vi)

94 What causes the leaf pores to open?
A high osmotic pressure in guard cell
B low osmotic pressure in guard cell
C turgidity of surrounding epidermis cells
D flaccidity of surrounding epidermis cells.

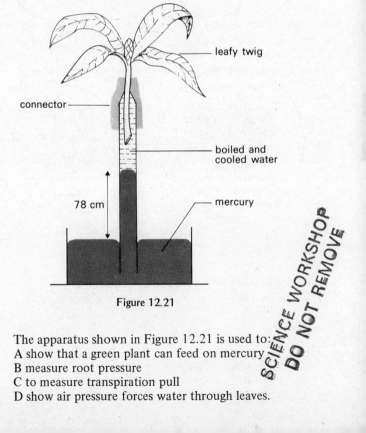

leafy twig

connector

78 cm

boiled and cooled water

mercury

Figure 12.21

95 The apparatus shown in Figure 12.21 is used to:
A show that a green plant can feed on mercury
B measure root pressure
C to measure transpiration pull
D show air pressure forces water through leaves.

96   Which one of the following plants is an example of an hydrophyte?

A heather   C buttercup
B water-lily   D tulip.

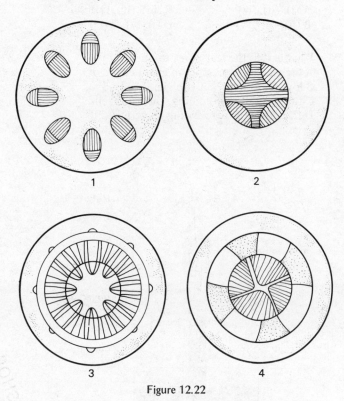

Figure 12.22

97   Which one of the general sectional view diagrams shown in Figure 12.22 represents a two-year-old twig?

A 1   C 3
B 2   D 4.

98   An example of a meristem tissue is the:

A xylem   C cambium
B phloem   D cortex.

99   Meristem tissue is composed of cells which are:
A able to divide and produce cells different to themselves
B made of woody material and carry water and salts
C soft cells carrying sugars and manufactured foods
D thin-walled cells with large vacuoles containing food reserves.

Questions 100–104 refer to Figure 12.23.

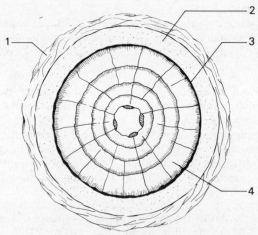

Figure 12.23

100   What is the approximate age of the woody stem?

A 2 years   C 10 years
B 5 years   D 20 years.

101   Which labelled part represents the spring and summer growth of the stem?

A 1   C 3
B 2   D 4.

102   Cork cells will be found in part labelled:

A 1   C 2
B 4   D 3.

103   In order to 'ring' the stem (see Question 85), it is
       necessary to remove parts labelled:
       A 1 and 2   C 1, 4, and 2
       B 1 and 4   D 1, 4, 2, and 3.

104   When a plant is propagated by grafting which one of
       the tissues, common to both the scion and stock, join
       together?
       A 1   C 2
       B 4   D 3.

105   Which two parts of a flowering plant provide the
       largest total surface area?
       (i) root hairs, (ii) stems, (iii) leaf blades, (iv) branch
       roots, (v) leaf stalks, (vi) bud scales.
       A (i), (iii)   C (iii), (v)
       B (ii), (iv)   D (iv), (vi)

106   The name given to a root which develops from a
       stem cutting is:
       A lateral   C adventitious
       B tap   D contractile.

13

Plant Nutrition

Questions 1–4 refer to the following experiment and Figure 13.1. A mixture of dry copper oxide and dry sugar from sugar beet plants was heated together in the test-tube.

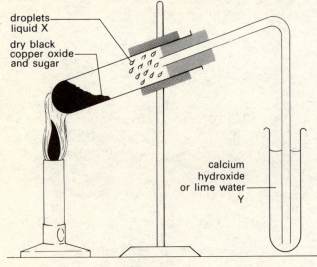

droplets liquid X

dry black copper oxide and sugar

calcium hydroxide or lime water Y

Figure 13.1

A gas was produced which turned the clear lime water milky or cloudy and droplets of liquid collected in the cooler parts of the test-tube.

1   If the droplets of liquid X turned white anhydrous copper sulphate a blue colour, the liquid X would be:
A ammonia   C hydrogen sulphide
B water   D carbon dioxide.

2   The liquid X was originally obtained from one of the following sources by the growing sugar beet plant.
A air   C nectaries
B soil   D sea water.

3   The gas produced from the beet sugar on heating it with dry copper oxide which turns lime water milky is:
A oxygen   C carbon dioxide
B ammonia   D carbon monoxide.

4   The gas referred to in Question 3 was originally obtained from one of the following sources by the growing sugar beet plant.
A air   C nectaries
B soil   D limestone.

5   Glucose, sucrose, and starch are each examples of chemical substances called:
A lipids   C proteins
B carbohydrates   D vitamins.

6   Which one of the following food substances would produce a blue-black colour when treated with iodine solution?
A glucose   C starch
B sucrose   D albumen.

7   In order to test chemically for glucose one of the following reagents is used:
A iodine solution   C Millon's solution
B Benedict's solution   D nitric acid.

8   Two of the following are parts of the green flowering
plant where food is manufactured.
(i) bark, (ii) green stems, (iii) root, (iv) floral petals,
(v) foliage leaves, (vi) fruits.
A (i), (vi)   C (iii), (iv)
B (ii), (v)   D (iv), (vi)

9   Green plants which can manufacture their own food
from simple raw materials have a method of nutrition
which is called:
A holophytic   C parasitic
B holozoic   D saprophytic.

10 Which three of the following are raw materials required
for food manufacture by green plants:
(i) proteins, (ii) carbon dioxide, (iii) carbohydrates,
(iv) water, (v) oxygen, (vi) minerals.
A (i), (iii), (v)   C (ii), (iii), (v)
B (ii), (iv), (vi)   D (i), (iv), (vi)

11 The process of food manufacture by green plants is
called:
A photosynthesis   C phototropism
B photoperiodism   D phototaxism.

12 Which one of the following can perform photosynthesis
and phototaxis?
A paramoecium   C sweet pea flowering
   protozoa   plant
B chlamydomonas algae   D earthworm.

13 In which labelled part of Figure 13.2 showing leaf
structure does the main process of food manufacture
take place?
A 1   C 4
B 3   D 2.

14 The supply of carbon dioxide to the food manufacturing
cells in Figure 13.2 can come from part labelled:
A 3   C 1
B 4   D 2.

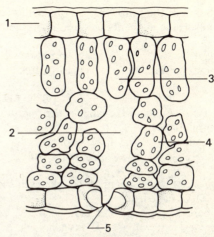

Figure 13.2

15 Stomata can restrict the supply of one of the following
raw materials for food manufacture:
A carbon dioxide   C oxygen
B water   D mineral salts.

16 If the stomata are closed the raw material prevented
from entering the leaf for food manufacture can be
provided alternatively by the:
A epidermis cells   C mitochondria
B cuticle   D chloroplasts.

17 The organelles concerned with the main process of food
manufacture in flowering plants are called:
A mitochondria   C chloroplasts
B nuclei   D companion cells.

18 An aquatic plant such as Canadian pondweed or
spirogyra can obtain a raw material for food
manufacture, which most other land plants obtain from
the air, from one of the following compounds dissolved
in water:
A sodium bicarbonate   C sodium chloride
B sodium nitrate   D sodium phosphate.

19   Which one of the following cell organelles contain the light-absorbing pigments for food manufacture in plants?
A nuclei   C chloroplasts
B mitochondria   D erythrocytes.

20    $$6CO_2 + 6H_2O \longrightarrow C_6H_{12}O_6 + 6O_2$$
What type of chemical change is the one shown in the equation above?
A neutralization   C synthesis
B hydrolysis   D condensation.

21   The product of the chemical change with the formula $C_6H_{12}O_6$ is called:
A glucose   C starch
B sucrose   D oxygen.

22   The gaseous by-product of food manufacture in green plants passes from the leaf cells to reach the surrounding air by the process of:
A osmosis   C capillarity
B diffusion   D radiation.

23   A specially prepared reagent called bicarbonate indicator will change its colour in the presence of carbon dioxide as one of the following:
A purple to yellow   C red to blue
B yellow to purple   D blue to red.

24   Carbon dioxide is a chemical compound which is:
A strongly alkaline   C weakly acid
B strongly acid   D weakly alkaline.

25   The correct path taken by water to reach the plant part where photosynthesis occurs is one of the following:
A stomata – air space – spongy mesophyll – palisade – companion cell
B root hair – xylem – phloem – mesophyll – mitochondria
C stomata – cortex – phloem – pith – nucleus
D root hair – cortex – xylem – mesophyll – chloroplast.

26   Which one of the following fine structures of the palisade chloroplast contains the green pigment?
A grana   C stroma
B membrane   D cellulose wall.

27   The green pigment chlorophyll is a:
A single chemical element
B single chemical compound
C mixture of chemical elements
D mixture of chemical compounds.

28   Fallen autumn leaves which contain large amounts of tannins have one of the following predominant colours:
A green   C brown
B yellow   D orange.

29   Chlorophyl has a colour composition made up of:
A green and yellow pigments
B green, yellow, and orange pigments
C green, yellow, and brown pigments
D green, yellow, and blue pigments.

30   Green leaves dipped in boiling water, then soaked in hot ethanol (ethyl alcohol), and then dipped in iodine solution and washed in cold water, appear to have a blue-black colour. This test indicates a leaf contains:
A glucose   C starch
B sucrose   D protein.

31   The hot ethanol (ethyl alcohol) used in the test in Question 30 will remove one of the following from green leaves:
A chlorophyll   C glucose
B starch   D carbon dioxide.

32   When green leaves are ground with clean sand and water in a mortar and pestle and the liquid extract filtered, the filtrate then heated with Benedict's solution produces a greenish-yellow to brown coloration. This test indicates a leaf contains:
A chlorophyll   C glucose
B starch   D ethanol (ethyl alcohol).

33   Which one of the following green leaves would give a
     positive result with the test used in Question 32?
     A cabbage   C geranium
     B iris   D beech.

34   Which one of the following statements is correct
     concerning the first-formed product of photosynthesis
     in leaves?
     A starch changes rapidly to sucrose
     B glucose changes rapidly to starch
     C sucrose changes rapidly to glucose
     D starch changes rapidly to glucose.

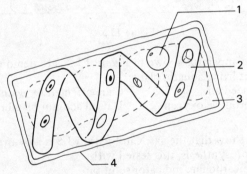

Figure 13.3

35   The spirogyra cell shown in Figure 13.3 has been
     growing in bright sunlight and is then placed in a
     solution of iodine and the cell viewed through a micro-
     scope. Which labelled part of the cell will be coloured
     blue-black?
     A 1   C 3
     B 2   D 4.

36   Which one of the following plants, after exposure to
     growth in sunlight, would produce a blue-black colour
     when treated with iodine solution?
     A mucor   C mushroom
     B yeast   D chlamydomonas.

37   A varigated leaf of laurel or pelargonium lacks one of
     the following in the white or non-green parts of the
     leaf:
     A palisade   C stomata
     B spongy mesophyll   D chlorophyll.

38   Clear colourless acetone after being ground up with
     dried green leaves separates as a clear green extract. A
     drop of the green coloured extract placed on a clean
     filter paper results in a green spot surrounded by a
     yellow ring. Which of the following have a yellow colour?
     A chlorophyll A   C xanthophyll
     B chlorophyll B   D carotene.

39   Flowering plants having all the essential nutrients but
     which have been grown in the dark for a long time,
     become pale yellow in colour and are called:
     A chlorotic   C phototropic
     B etiolated   D dwarfed.

40   The plants in Question 39 are deficient in glucose and
     starch and also one of the following:
     A chlorophyll   C magnesium
     B iron   D vitamin D.

41   A disc X is punched out of a leaf from a growing plant,
     then dried at 110 °C and weighed. A week later another
     disc Y is punched out of the leaf of the same growing
     plant using the same punch, dried at 110 °C and weighed.
     Leaf disc Y is found to have a greater dry weight than
     leaf disc X; this is because leaf Y has:
     A increased in area   C produced more carbohydrate
     B more water in it   D transpired at a greater rate.

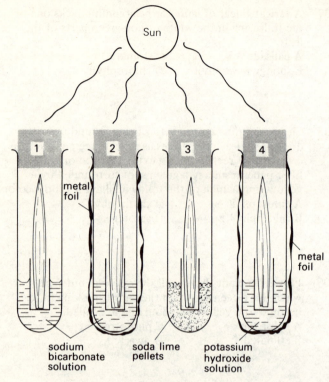

Figure 13.4

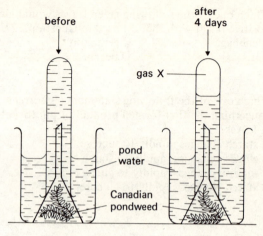

Figure 13.5

42 Each tube in Figure 13.4 contains an *etiolated* iris leaf in a small cup of nutrient solution. Tubes 2 and 4 are blacked out with metal foil. Tubes 1 and 2 contain sodium bicarbonate, whilst tube 3 has soda lime pellets, and tube 4 has potassium hydroxide solution. After one hour the leaves are tested for glucose which is found to be present in *one* of the following:

A 2   C 4
B 3   D 1.

43 Which tube in Figure 13.4 was deprived of *more than one* essential requirement for photosynthesis to occur?

A 2   C 4
B 3   D 1.

44 The apparatus shown in Figure 13.5 is to demonstrate that living aquatic plants:

A need air   C are affected by water pressure
B produce oxygen   D need sodium bicarbonate.

45 To show that the gas X in Figure 13.5 is a by-product of photosynthesis, it is tested with:

A bicarbonate indicator solution
B a glowing wood splint
C iodine solution
D lime water solution.

46 The gas X collecting in the tube shown in Figure 13.5 is:

A pure oxygen
B a mixture of oxygen and carbon dioxide
C air enriched with oxygen
D pure carbon dioxide.

47 Which one of the following gases could be of value to growing plants when introduced into the air of a greenhouse?

A ammonia   C carbon dioxide
B carbon monoxide   D sulphur dioxide.

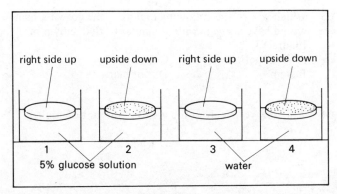

right side up   upside down   right side up   upside down

1   2   3   4

5% glucose solution   water

Figure 13.6

48 Discs of equal size cut from *etiolated* leaves of geranium are placed in a 5 per cent solution of glucose (see Figure 13.6), some are floating right side up and some are upside down. Other discs are placed in *water* some floating upside down and others right side up. All the leaf discs and solutions are then placed in total darkness for one week. Which one of the discs on testing with iodine solution will show the presence of most starch?
A 1   C 3
B 2   D 4.

49 The experiment in Question **48** shows that a stage of photosynthesis:
A takes place in the daylight
B takes place in the dark
C is controlled by osmosis
D changes carbohydrates into proteins.

50 Which one of the following is first formed during the sunlight stage of photosynthesis?
A sucrose   C glucose
B starch   D maltose.

51 Which one of the following is involved in the downward movement of the products of photosynthesis from the shoot system to the root?
A phloem   C cambium
B xylem   D cortex.

52 In what form do the carbohydrate products of photosynthesis travel in Question **51**?
A cellulose   C sucrose
B starch   D amino acids.

53 The process of transportation of dissolved substances within a plant is called:
A transpiration   C carbon assimilation
B translocation   D circulation.

54 The movement of dissolved substances within a plant can be investigated by:
A radioactive tracers   C incomplete culture solutions
B cobalt chloride papers D potometers.

55 If the bark of an apple tree is 'ringed' soon after petal fall, which of the following will be deprived of the products of photosynthesis?
A roots   C fruits
B stems   D leaves.

56 Which two of the following tissues remain attached to the plant after ringing a stem?
(i) cortex, (ii) pith, (iii) cambium, (iv) xylem, (v) phloem, (vi) epidermis.
A (i), (vi)   C (iii), (v)
B (ii), (iv)   D (v), (vi)

57 After flower formation the part of an annual broad-leaved flowering plant receiving most of the products of photosynthesis is the:
A root   C leaf
B fruit   D buds.

58 The part of a biennial plant receiving most of the products of photosynthesis towards the end of the first year of growth is the:
A root   C leaf
B fruit   D buds.

59 Which one of the following is a form of carbohydrate used in food storage organs of a plant?
A starch   C glucose
B pyrenoid   D aleurone.

60 Which one of the following carbohydrates is a plant structural material?
A starch   C chlorophyll
B cellulose   D glycogen.

61 The product of photosynthesis for immediate use to provide energy for young growing plants is:
A sucrose   C starch
B glucose   D chlorophyll.

Questions **62–64** refer to Figure 13.7, a graph obtained by plotting the sugar content in leaves and a certain organ of a plant at different times in one day.

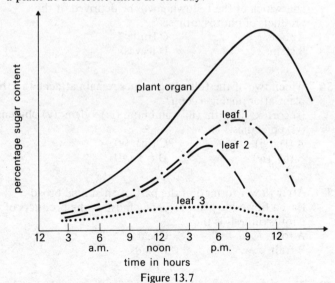

Figure 13.7

62 From which of the green leaves was light completely excluded?
A 1   C 3
B 2   D 1 and 2.

63 Which one of the following organs of the growing plant would have a sugar content similar to that shown in · Figure 13.7?
A a leaf petiole   C tap root
B stem   D root hair.

64 Which leaves received the greatest amount of sunlight?
A 1   C 3
B 2   D 1 and 2.

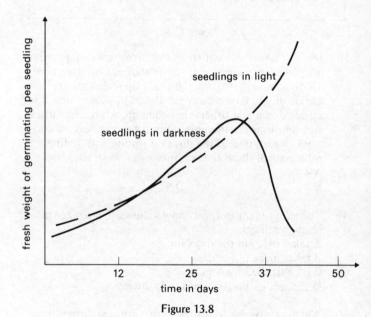

Figure 13.8

65 The graph in Figure 13.8 shows the change in fresh weight of germinating pea seedlings when one group was grown in complete darkness and the other group received regular daylight. At what approximate period of time did the light-receiving seedlings commence to produce their own carbohydrate by photosynthesis?
A 0–10 days   C 25–30 days
B 10–20 days   D 35–50 days.

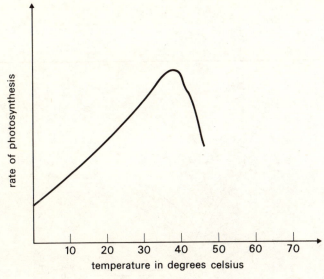

Figure 13.9

66 The graph in Figure 13.9 shows the affect of increasing temperature on the rate of photosynthesis of green leaves. At what approximate temperature did the process of photosynthesis stop?

A between 0–10 °C   C 30–40 °C

B 20–30 °C   D 40–60 °C.

67 The reason for the halting of photosynthesis with increasing temperature in Question **66** is that:

A leaves dry out   C leaves catch fire

B enzymes are killed   D glucose turns to caramel.

14

Plant Respiration and Growth

1 Which one of the following is the way by which air enters the leaf of a flowering plant?
A lenticels   C cork
B stomata   D cuticle.

2 The method of entry of air into green algae and fungi is by way of:
A stomata   C osmosis
B diffusion   D lenticels.

3 Which of the following is the correct path taken by oxygen during external respiration in a flowering plant?
A external air – stomata – mesophyll air-space – cell wall
B stomata – cell wall – external air – mesophyll air-space
C cell wall – external air – mesophyll air-space – stomata
D mesophyll air-space – cell wall – stomata – external air.

4 In which one of the following does the process of internal respiration take place in a flowering plant?
A only in leaf cells   C only in root cells
B only in stem cells   D in all plant cells.

5 Figure 14.1 shows a leafy shoot firmly fixed in a rubber stopper, and the cut end of the twig immersed in water. When the vacuum pump is on, a stream of bubbles is seen to come from the cut end of the stem. This experiment demonstrates that:
A air enters the leaves of a plant
B air enters the stem of a plant
C air spaces are continuous throughout a plant
D the root of a plant can give out carbon dioxide.

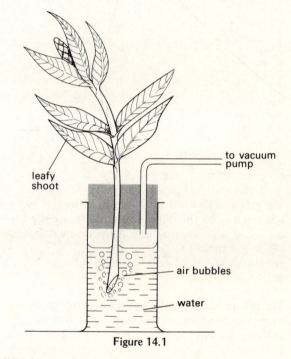

leafy shoot

to vacuum pump

air bubbles

water

Figure 14.1

6 Which one of the following is the respiratory centre of a plant cell?
A chloroplasts   C mitochondria
B chromosomes   D cell wall.

7 Which one of the following can provide a plant's respiratory centres with oxygen as an alternative to being supplied by the air?
A chloroplast   C starch grain
B nucleus   D pyrenoid.

8   Which of the following is the best definition of respiration in plants?
A movements of the atomata
B exchange of oxygen with carbon dioxide
C release of carbon dioxide and water
D release of energy from food.

9   One of the following flowering plant physiological processes could not take place in darkness:
A respiration   C transpiration
B photosynthesis   D absorption.

10 Which one of the following are capable of anaerobic respiration for limited periods?
A all plants   C fungi
B flowering plants   D algae.

11          $C_6H_{12}O_6 + 6O_2 \rightarrow 6CO_2 + 6H_2O$
This chemical change occurs during one of the following plant processes:
A anaerobic respiration   C photosynthesis
B aerobic respiration   D transpiration.

12 The chemical change indicated in Question 11 is one of the following type:
A reduction   C oxidation
B neutralization   D condensation.

13 Which one of the following has been left out of the equation in Question 11?
A chlorophyll   C kinetic energy
B sunlight   D potential energy.

14 The raw material needed for the plant process shown in Question 11, in addition to oxygen, is:
A glucose   C carbon dioxide
B starch   D water.

15 Which one of the following reagents can absorb oxygen?
A lime water   C potassium hydroxide
B potassium pyrogallate   D soda lime.

16 Which one of the following is a product of anaerobic respiration and not produced by aerobic respiration?
A carbon dioxide   C oxygen
B water   D ethanol.

17 The energy stored in glucose originally came from:
A electricity   C the air
B sunlight   D the soil.

18 Which one of the following can provide an *animal* with energy?
A water   C sodium chloride
B ethanol   D iodine.

19 Which three of the following are means of oxygen entry into a green flowering plant?
(i) cuticle, (ii) guard cells, (iii) root hairs, (iv) pith, (v) lenticels, (vi) xylem.
A (i), (iv), (vi)   C (i), (iii), (vi)
B (ii), (iii), (v)   D (ii), (iv), (v)

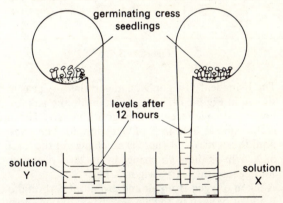

Figure 14.2

20 Equal amounts of germinating cress seedlings are placed in the retort flasks shown in Figure 14.2 and both maintained at a constant temperature in daylight for 12 hours. The ends of the retorts dip into two different liquids X and Y, the level of X rises whilst Y remains unchanged. Which of the following is solution X?
A olive oil   C potassium hydroxide
B saturated brine   D mercury.

21   What is the reason for the level of liquid X rising in
Figure 14.2?
A a fall in temperature causes contraction of air in the
retorts
B the carbon dioxide liberated by the seedlings is absorbed
C oxygen has been absorbed from the air
D water vapour is condensing on cool glass surfaces.

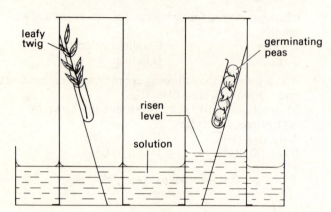

Figure 14.3

22   The living leafy twig in X, and germinating peas in Y
shown in Figure 14.3 have been inside the jars for 24
hours and have received light continuously. The liquid
is the same as used in X of Question 20. The level of the
liquid rises in Y but there is no change in the level
of X. Why is there *no* change in the level of liquid in
X? Because the leafy twig in X has:
A taken oxygen from air and given out carbon dioxide
B added oxygen to the air, no carbon dioxide given out
C given out carbon dioxide and water vapour
D given out carbon dioxide without taking in any gas.

23   Which one of the following gases will increase in amount,
or concentration, in the air containing the leafy twig X
in Question 22 compared to the air above the
germinating peas in Y?
A nitrogen   C carbon dioxide
B oxygen   D water vapour.

24   Which one of the following would produce the same
experimental results as the germinating peas in Y of
Question 22?
A geranium plant   C ripening apples
B dried peas   D lettuce leaves.

25   Which one of the following would produce the same
experimental results as the leafy twig X in Question 22?
A mushrooms   C slice of potato
B red rose petals   D variegated laurel leaves.

26   What is the normal carbon dioxide percentage content
of room air?
A 3   C 0.03
B 30   D 0.30.

27   Which one of the following is able to carry on the process
of photosynthesis?
A pear fruits   C germinating bean seeds
B cabbage leaves   D buttercup roots.

28   Which one of the following is *unable* to carry on the
process of respiration?
A rhubarb stems   C dried raisins
B buttercup petals   D tomato fruits.

29   In which tube shown in Figure 14.4 would the soaked
pea seeds be unable to germinate?
A 1   C 3
B 2   D 4.

30   The graph in Figure 14.5 shows the rate of respiration
during the complete life cycle of a garden pea plant.
Which part of the graph corresponds to the rate of
respiration of a 14-day-old germinating pea seedling?
A 1   C 3
B 2   D 4.

31   When apples are kept in storage one of the following
gases in the stored air slows down the rate of ripening
and is kept at a certain concentration.
A water vapour   C oxygen
B carbon dioxide   D nitrogen.

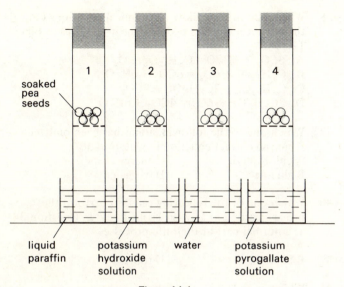

soaked pea seeds

1   2   3   4

liquid paraffin   potassium hydroxide solution   water   potassium pyrogallate solution

Figure 14.4

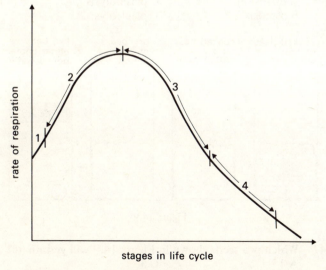

rate of respiration

stages in life cycle

Figure 14.5

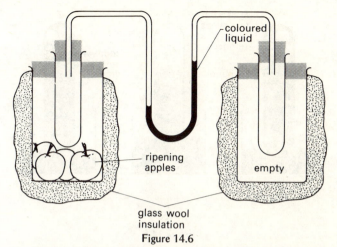

coloured liquid

ripening apples

empty

glass wool insulation

Figure 14.6

32   Figure 14.6 is an apparatus used for experiments with respiring plants. What is the object of the experiment using the ripening apples? The experiment shows that respiring plants:

A produce oxygen   C take in heat
B produce carbon   D give out heat.
  dioxide

fresh field mushroom   iron wire gauze   coloured water droplet movement

soda lime grains   Figure 14.7

33   The apparatus shown in the Figure 14.7, containing a fresh field mushroom, is used to show that non-green plants:

A produce heat during respiration
B require water for photosynthesis
C produce carbon dioxide during respiration
D require the mineral iron for healthy growth.

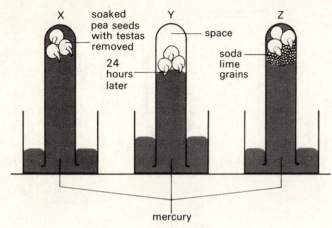

Figure 14.8

34 In Figure 14.8 soaked peas, with testas removed, are completely surrounded by mercury in X; 24 hours later the formation of a space is seen around the peas shown in Y. The space, appearing around the peas after 24 hours is due to:
A a vacuum   C ethanol (ethyl alcohol)
B oxygen   D carbon dioxide.

35 The apparatus in Figure 14.8 shows that pea seeds in the absence of air:
A can produce ethanol   C produce carbon dioxide
B produce oxygen   D create a low pressure.

36 Pieces of soda lime introduced into the space around the seeds in Y of Figure 14.8 cause the mercury level to rise rapidly and the space disappears, as in Z. This shows the space around the seeds contained:
A water   C oxygen
B carbon dioxide   D ammonia.

37 The soda lime test of Question **36** shows that the gas is a by-product of:
A aerobic respiration   C photosynthesis
B anaerobic respiration   D transpiration.

38 Which one of the following chemical equations indicate the chemical changes taking place in the peas in Figure 14.8?
A. $6CO_2 + 6H_2O = C_6H_{12}O_6 + O_2$
B. $C_6H_{12}O_6 + 6O_2 = 6CO_2 + 6H_2O$
C. $C_6H_{12}O_6 = 2C_2H_5OH + 2CO_2$
D. $2C_2H_5OH + 6O_2 = 4CO_2 + 6H_2O.$

39 Which one of the following items has been omitted from the correct equation in Question **38**?
A chlorophyll   C energy
B glucose   D oxygen.

40 Which one of the following can also perform the process carried out by the soaked peas in Question **34**, as a normal part of their life processes?
A green leaves   C fruits
B bacteria   D roots.

41 Which one of the following physical processes account for the movement of gases into a plant during respiration?
A diffusion   C plasmolysis
B osmosis   D electrolysis.

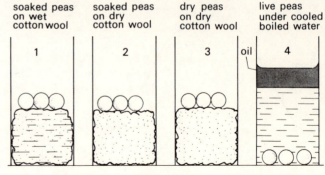

Figure 14.9

42 Which pea seeds shown in Figure 14.9 will germinate?
A 1   C 3
B 2   D 4.

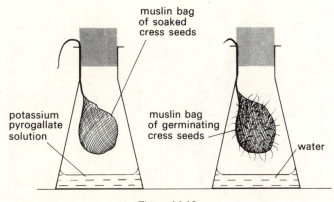

muslin bag
of soaked
cress seeds

potassium
pyrogallate
solution

muslin bag
of germinating
cress seeds

water

Figure 14.10

43   Figure 14.10 shows the appearance after 1 week of
two flasks which originally contained muslin bags of
soaked cress seeds. The experiments show that for seed
germination to take place:
A water is necessary   C oxygen is required
B oxygen is not required D a suitable temperature is
                             necessary.

44   The energy required for the early rapid growth during
the germination of broad bean seeds comes from:
A starch stored in the radicle
B chlorophyll present in the plumule
C glucose from the cotyledon
D minerals absorbed by the micropyle.

45   Which one of the following is an excretory waste
product of flowering plants?
A urea   C ammonia
B sodium chloride   D tannins.

46   Which two of the following are considered as
excretory waste materials in flowering plants?
(i) starch, (ii) cellulose, (iii) pine resin, (iv) aleurone
grains, (v) calcium oxalate, (vi) chlorophyll.
A (i), (iv)   C (iii), (v)
B (ii), (vi)   D (iv), (vi)

47   Which one of the following plants can eliminate waste
by defaecation?
A orchid   C seaweed
B sundew   D fern.

48   Which of the following plant materials is defaecated
by humans?
A glucose   C protein
B cellulose   D starch.

49   Which two parts of a secondary thickened tree trunk
contains most excretory waste?
(i) sapwood, (ii) heartwood, (iii) bark, (iv) phloem,
(v) cambium, (vi) xylem.
A (i), (iv)   C (iv), (v)
B (ii), (iii)   D (v), (vi)

50   Which one of the following plant products provide
useful sources of many medical drugs for man?
A gums   C auxins
B carbohydrates   D alkaloids.

51   In which part of a plant cell does the process of
mitosis take place?
A chloroplast   C cytoplasm
B vacuole   D nucleus.

52   In which part of a green flowering plant body does
the process of meiosis take place?
A root   C leaf
B stem   D flower.

53   One of the following is actively concerned with
division of a cell nucleus:
A mitochondria   C chromosomes
B chloroplasts   D ribosomes.

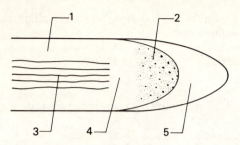

Figure 14.11

Questions **54–58** refer to Figure 14.11.

54   Figure 14.11 which shows a longitudinal section of the general appearance of the root tip of a broad bean seen through a microscope. Which part of the root consists mainly of actively dividing cells?
A 5   C 1
B 2   D 3.

55   In which part is the region of elongation found?
A 3   C 5
B 4   D 2.

56   The cells in the elongating region develop a large:
A nucleus   C chloroplast
B vacuole   D pyrenoid.

57   The region of differentiation in Figure 14.11 is the part labelled:
A 5   C 1
B 2   D 3.

58   The region of differentiation in Figure 14.11, will contain actively dividing cells in a tissue called the:
A root cap   C xylem
B cambium   D root hairs.

59   Where are the growth regions in a vertebrate animal?
A spinal column   C the head
B the legs   D in all cells.

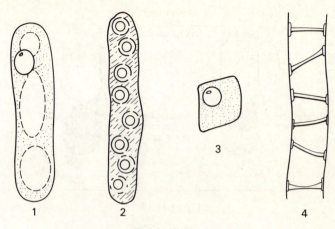

Figure 14.12

60   Which one of the cells shown in Figure 14.12 came from a root meristem?
A 1   C 3
B 2   D 4.

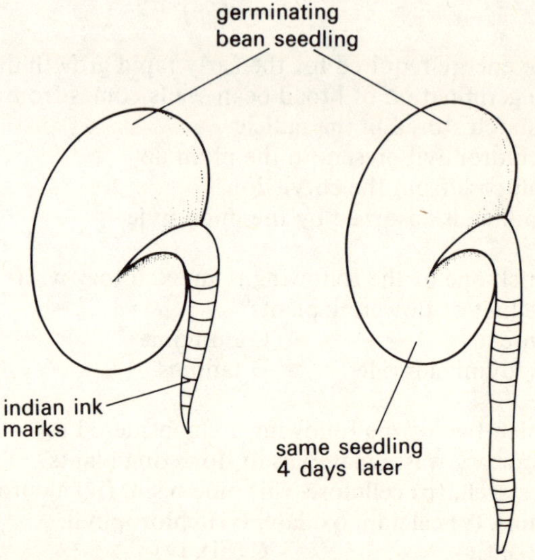

germinating
bean seedling

indian ink
marks

same seedling
4 days later

Figure 14.13

61 In Figure 14.13 the experiment is to show:
A the rate of growth in a bean seedling
B the region of maximum elongation in a root
C the radicle cannot absorb indian ink
D the root responds to gravity.

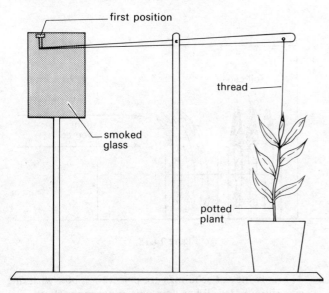

Figure 14.14

62 The apparatus connected to the potted plant shown in Figure 14.14 is to measure a plant's:
A rate of photo-synthesis
C transpiration pull
B increase in length
D response to gravity.

63 Which one of the following experimental methods or apparatus, repeated at intervals with different plants, would show the rate of growth of young wheat plants?
A the auxanometer apparatus
B heating the complete plant to 110° C and weighing to find dry weight
C recording weight of plant in pot covered with rubber sheet
D the potometer apparatus.

64 The increase in dry weight of one of the following will show the rate of photosynthesis in a piece of:
A apple
C cabbage leaf
B potato tuber
D carrot root.

65 The increase in dry weight of all the specimens in Question **64** will be a measure of the rate of:
A growth
C respiration
B transpiration
D translocation.

66 A growing green plant is found to show a decrease in its dry weight when subjected to one of the following growing conditions:
A continually in artificial light
B continually in darkness
C partly in daylight (75 per cent) and darkness (25 per cent)
D partly in daylight (50 per cent) in darkness (50 per cent).

67 Which one of the following growth requirements can be altered or provided by a farmer in order to increase the wheat crop?
A sunlight
C carbon dioxide
B mineral salts
D temperature.

68 If a green plant is growing in complete darkness which one of the following processes is it *unable* to perform?
A transpiration
C photosynthesis
B respiration
D absorption.

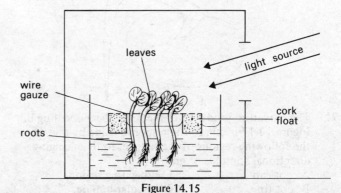

Figure 14.15

69 As shown in Figure 14.15, cress seeds are allowed to germinate on a floating raft in a box with one sided illumination. The experiment shows that the curvature produced in the *roots* is due to:

A negative geotropism   C negative phototropism
B positive hydrotropism   D positive geotropism.

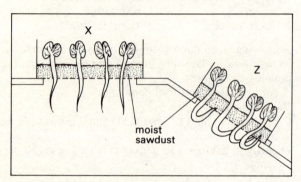

Figure 14.16

70 The cress seeds shown in Figure 14.16 are allowed to germinate in complete darkness on moist sawdust supported on sieves. The resulting curvature of the cress roots in Z, is due to:

A negative geotropism   C positive hydrotropism
B positive phototropism   D negative phototropism.

Figure 14.17

71 The curvature produced in the broad bean seedling in Figure 14.17 is due to the response made by one of the following regions of the radicle, to the changing directional stimulus of gravity:

A root cap   C elongating region
B root tip   D vascular bundle.

72 The unicellular algae, chlamydomonas, can move towards the source of one-sided illumination. Such a plant movement is called:

A positive phototropism   C negative photonastism
B positive phototaxism   D photosynthesis.

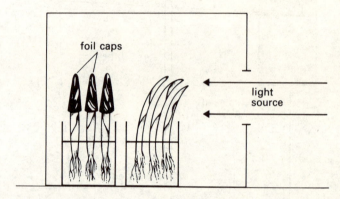

Figure 14.18

73 The growing oat seedlings are placed in a box with one side illuminated. What does this experiment show concerning the effect of light on the growth of the seedlings with foil caps on the shoot tips, compared to those without? The foil-capped shoot:

A is too far away from the light source to be stimulated
B is deprived of oxygen for growth and cannot produce a curvature
C without light cannot photosynthesise, therefore no growth curvature
D tips must receive the stimulus.

74 Rhizomes and runners are modified stems. What type of response do they show to gravity?

A positive geotropic   C diageotropic
B negative geotropic   D positively hydrotropic.

75   Which one of the following would account for the
increase in size of a plant leaf during growth?
A increase of cell water content
B mitotic division of cells
C increase of cell food reserves
D meiotic division of cells.

76   In which one of the following plant organs will the
process of meiosis take place?
A leaf   C root
B stem   D flowers.

77   Which one of the following will show meiosis?
A sepals   C anthers
B petals   D floral axis.

15

Plant Reproduction

1 A group of flowers found on one main stem is called:
   A a floret
   B an inflorescence
   C a floral axis
   D a bouquet.

2 One of the following protects a flower when it is in bud:
   A calyx
   B corolla
   C androecium
   D gynaecium.

3 Which one of the following flower parts consists of filament and anthers?
   A calyx
   B corolla
   C androecium
   D gynaecium.

4 Plants forming male and female flowers in separate plants are called:
   A hermaphrodite
   B monoecious
   C dioecious
   D polygamous.

5 Which one of the following are components of the flower perianth?
   A sepals/petals
   B petals/stamens
   C sepals/pistil
   D stamens/pistil.

6 Scent glands are normally found in one of the following:
   A nectaries
   B petals
   C filaments
   D receptacles.

7 Which one of the following is removed from strawberry fruits before eating?
   A petals
   B sepals
   C stamens
   D pistils.

8 Which one of the following has composite flowers?
   A buttercup
   B tulip
   C daisy
   D sweet pea.

9 Which one of the following have petals arranged in a crosswise manner within the flower?
   A rose
   B daffodil
   C wallflower
   D dandelion.

10 Pollen grains are formed in large numbers within one of the following:
    A basket
    B tube
    C sacs
    D coat.

11 A pollen grain is considered to be one of the following:
    A male gamete
    B female gamete
    C spore
    D zygote.

12 Which one of the following is contained within the carpel of a flower?
    A pollen grain
    B oviduct
    C sperm
    D ovules.

Questions 13 and 14 refer to Figure 15.1.

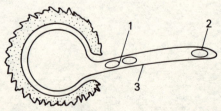

Figure 15.1

13 The two nuclei, part labelled 1, are called:
A spermatozoa   C tube nuclei
B male gametes   D ovules.

14 The part labelled 3 is called a:
A filament   C conjugation canal
B pollen tube   D sporogonium.

15 Pollen grains when added to 10 per cent sucrose solution produce structures after 24–48 hours. This process taking place in the sucrose solution is called:
A pollination   C germination
B fertilization   D spawning.

16 Which two of the following plants produces very light, small and smooth-coated pollen grains?
(i) rose, (ii) tulip, (iii) willow, (iv) hazel, (v) buttercup, (vi) sweet pea.
A (i), (ii)   C (iii), (iv)
B (ii), (v)   D (v), (vi)

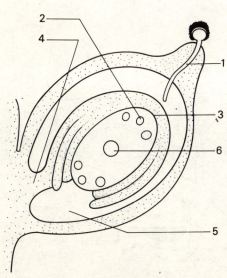

Figure 15.2

17 Figure 15.2 shows a flower gynaecium together with an attached pollen grain. Which labelled part is the egg nucleus?
A 2   C 3
B 5   D 6.

18 The pollen cell extension 1 is shown passing through one of the following to reach the egg cell:
A oviduct   C style
B micropyle   D filament.

19 If the chromosome number of the mature plant gynaecium shown in Figure 15.2 is 16, what will be the number of chromosomes in each of the egg nucleus and male gamete?
A 8   C 24
B 16   D 32.

20 Which one of the following is a feature typical of a grass inflorescence?
A large petals   C large nectaries
B feathery stigmas   D well scented.

21 The state of the chromosomes in a normal zygote nucleus is said to be:
A haploid   C tetraploid
B diploid   D polyploid.

22 The process of fusion between the male gamete and the egg cell in a flower is called:
A pollination   C gametogenesis
B fertilization   D germination.

23 After the process of zygote formation in a tomato flower, which one of the following does not entirely wither and remains attached to the fruit?
A petals   C sepals
B anthers   D filament.

24   Which three of the following make up the *embryo* of a seed?
(i) cotyledon, (ii) endosperm, (iii) testa, (iv) radicle, (v) pericarp, (vi) plumule.
A (i), (ii), (iii)   C (ii), (v), (vi)
B (i), (iv), (vi)   D (iii), (iv), (v)

25   Which one of the following is an acceptable definition of pollination?
A transfer of pollen from stigma to anther
B transfer of pollen from anther to stigma
C union of male and female zygote to form gametes
D union of male and female gamete to form zygotes.

26   A flower is said to be protandrous when:
A it has male flowers only
B it has female flowers only
C the carpel ripen before stamens
D the stamens ripen before carpels.

27   The term which is opposite to protandrous is:
A perigynous   C protogynous
B epigynous   D hypogynous.

28   A protandrous flower is unable to be pollinated by:
A wind   C water
B itself   D animal.

29   Which one of the following insects have a long tongue?
A butterfly   C cockroach
B wood-louse   D ant.

30   Which one of the following provide the hive bee with protein food?
A nectaries   C scent glands
B anthers   D ovules.

31   Which one of the following plant flowers are visited by insects seeking food?
A hazel catkins   C apple
B barley   D oats.

Figure 15.3

32   The type of flower shown in Figure 15.3 is pollinated by one of the following agents:
A water   C insects
B wind   D slugs.

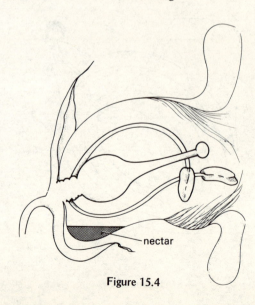

nectar

Figure 15.4

33   The sectional view of a snapdragon-type flower shown in Figure 15.4 can only be pollinated by:
A ants   C wind
B bees   D birds.

34   Which four of the following are features of *insect* pollinated flowers?
(i) smooth pollen, (ii) rough pollen, (iii) feathery stigma, (iv) sticky stigma, (v) scented, (vi) scentless, (vii) coloured corolla, (viii) nectarless.
A (i), (iv), (v), (viii)   C (i), (iii), (vi), (vii)
B (ii), (iv), (v), (vii)   D (ii), (iii), (vi), (viii)

35   Which one of the following food producers would wisely accept beehives next to their growing crop?
A mushroom grower   C rose grower
B pear grower   D wheat farmer.

36   Cross pollination results in one of the following:
A greater variation of offspring
B weaker offspring
C fewer offspring
D no variation of offspring.

37   Which one of the following methods would be used to maintain a special variety of sweet pea?
A cross pollination   C root cuttings
B self pollination   D bulbs.

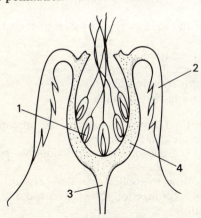

Figure 15.5

38   Figure 15.5 shows a transverse sectional view of the structure of a rose hip. Which labelled part is called the peduncle?
A 3   C 4
B 2   D 1.

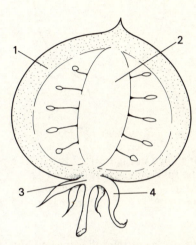

Figure 15.6

39   Questions 39–42 refer to Figure 15.6 which is a transverse sectional view of a tomato fruit. Which part of the fruit is the floral axis or receptacle?
A 1   C 2
B 3   D 4.

40   The type of fruit shown in Figure 15.6 is called a:
A juicy fruit   C false fruit
B dry fruit   D box fruit.

41   The central part labelled 2 is the:
A pericarp   C placenta
B testa   D seed.

42   The tomato fruit is an example of one of these:
A drupe   C pome
B berry   D capsule.

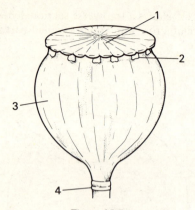

Figure 15.7

43 The fruit shown in Figure 15.7 is that of a poppy. Which one of the following is part labelled 1?
A petal scar   C pericarp
B anthers   D stigmas.

44 The fruits of one of the following can burst with explosive force to disperse their seeds:
A gorse   C dandelion
B gooseberry   D horse-chestnut.

45 Which part of the fruit splits in Question **44**?
A testa   C mesocarp
B pericarp   D pappus.

46 Which one of the following is a component part of a fruit wall?
A micropyle   C pericarp
B plumule   D cotyledon.

47 One of the following produce fruits in which the testa is a component part of the fruit wall:
A legumes   C cereals
B pomes   D marrows.

48 All true seeds can be distinguished from fruits by the presence of one of the following:
A stigma   C micropyle
B petal scar   D endosperm.

49 Which one of the following is a feature of the seeds of one group of flowering plants?
A plumule   C one cotyledon
B radicle   D testa.

50 Which one of the following is a function of the cotyledon of dwarf French beans and is *not* a function of the broad bean cotyledon?
A food storage in the seed
B protect plumule in the seed
C to function as first leaves
D to swell with water.

51 Which one of the following seeds shows hypogeal germination?
A sunflower   C cress
B castor   D pea.

52 Gorse, sweet pea, lupin, and broom all produce fruits which are:
A capsules   C pods
B nuts   D berries.

53 Which of the following is a process of sexual reproduction in plants?
A spore formation in fern
B spore formation in moulds
C seed formation in potato
D tuber formation in dahlia.

54 Which of the following is the correct representation of the life cycle of a flowering plant?
A thallus — archegonium — antheridium — sporangium — spore
B seed — flower — pollen — ovule — zygote
C thallus — conjugation tube — zygospore — spores — prothallus
D seed — cone — pollen — ovule — zygote.

55 Which four of the following are features possessed by dicotyledon flowering plants?
(i) one seed leaf, (ii) two seed leaves, (iii) narrow leaves, (iv) broad leaves, (v) net veins, (vi) parallel veins, (vii) perianth in threes, (viii) perianth in fives.

A (i), (iii), (vi), (vii)   C (i), (iv), (v), (vii)
B (ii), (iv), (v), (viii)   D (ii), (iii), (vi), (viii)

56 Which two of the following are monocotyledon flowering plants?
(i) garden pea, (ii) daffodil, (iii) rose, (iv) wheat, (v) maize, (vi) hazel.

A (i), (ii), (vi)   C (ii), (iv), (v)
B (i), (iii), (iv)   D (iii), (v), (vi)

57 Which one of the following can be reproduced by a method other than by producing seeds?
A sweet pea   C apple
B broad bean   D sunflower.

58 Which one of the following plants is propagated vegetatively?
A spring cabbage – 'Greyhound'
B rose – 'Peace'
C cos lettuce
D peas – 'Kelvedon Wonder'.

59 Which of the following members of a family have been partly formed asexually?
A boy   C twin boy and girl
B girl   D identical twin boys.

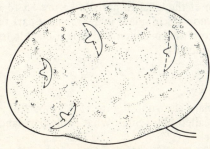

Figure 15.8

60 Figure 15.8 shows a potato tuber. Which one of the following indicate the plant organ from which it is derived?
A has lateral roots with root hairs
B has many swollen bud leaves
C has many petal and sepal scars
D has many buds in scale leaves.

61 What is the maximum number of daughter plants that could be obtained by cutting up the potato tuber shown in Figure 15.8? (All structural features are shown in the diagram.)
A 1   C 4
B 2   D 5.

62 What type of fruits are produced by potato plant flowers?
A nuts   C dry pod
B juicy berry   D false fruit.

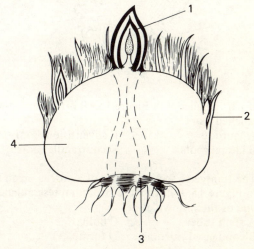

Figure 15.9

63 Figure 15.9 is the vegetive reproductive organ of a spring flower. What is the vegetative organ called?
A corm   C stem tuber
B bulb   D rhizome.

64 Where will the next year's daughter plant develop in Figure 15.9?

A 3	C 2
B 4	D 1.

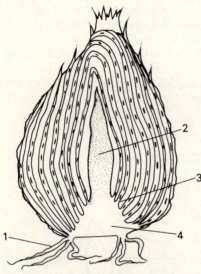

Figure 15.10

65 If part labelled 4 in Figure 15.10 is the stem, the parts labelled 1 are:

A taproots	C adventitious roots
B lateral roots	D contractile roots.

66 The transverse sectional view of a crocus organ shown in Figure 15.9, has a close structural resemblance to *one* of the following:

A stem tuber	C bulb
B condensed rhizome	D turnip.

67 What is the maximum number of new daughter plants that could be formed from the one shown in Figure 15.9?

A 1	C 3
B 2	D 4.

68 Which one of the following plants can be purchased as corms?

A tulip	C gladiolus
B daffodil	D onion.

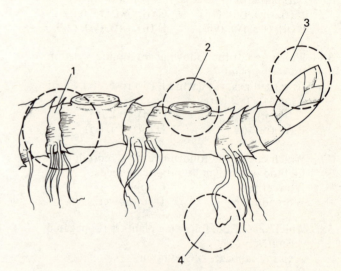

Figure 15.11

69 Figure 15.11 shows the underground portions of a 'Solomon's Seal' plant. This vegetative organ is called a:

A runner	C rhizome
B root stock	D root tuber.

70 Which of the labelled parts in Figure 15.11 of the 'Solomon's Seal' plant is the *smallest* part that could be cut from the specimen and would be able to survive and reproduce itself?

A 2	C 1
B 3	D 4.

71 The plant shown in Figure 15.12 is part of a cultivated rose bush called 'Shot Silk'. What is the part labelled X called?

A stock	C sucker
B scion	D runner.

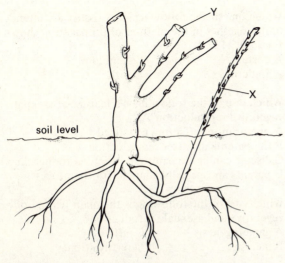

Figure 15.12

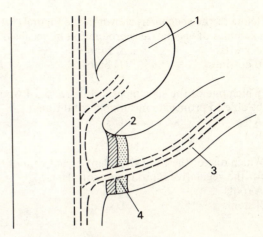

Figure 15.13

72 What type of flowers will be formed on X shown in Figure 15.12?

A lemon yellow   C wild rose

B deep purple   D scarlet red.

73 Strawberry plants called 'Royal Sovereign' are best propagated from:

A rhizomes   C seeds

B stools   D runners.

74 The most useful feature of vegetative reproduction is that the offspring:

A are strong, showing renewed vitality

B are free from parentally transmitted disease

C show great variation in flower colour etc.

D are identical to the parent plants.

75 Figure 15.13 shows a longitudinal sectional view of the tissues of a stem. Part 3 is a leaf petiole. The layer of cells labelled 4 are concerned with leaf fall and is called the :

A cork layer   C abscission layer

B piliferous layer   D cortical layer.

76 The part labelled 1 in Figure 15.13 is called:

A a terminal bud   C a bud scar

B an axilary bud   D a bulb.

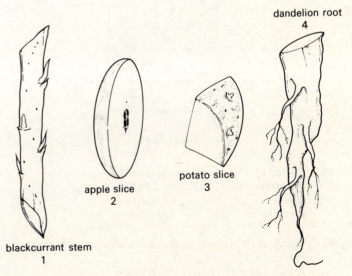

Figure 15.14

77   Three of the specimens shown in the Figure 15.14 are
      examples of vegetative reproduction by means of:
      A grafts   C runners
      B cuttings   D bulbs.

78   Which one of the specimens in Figure 15.14 would not
      be able to propagate in suitable conditions?
      A 1   C 3
      B 2   D 4.

79   Which one of the following plant organs cannot be used
      for the vegetative reproduction of plants?
      A roots   C leaves
      B stems   D seeds.

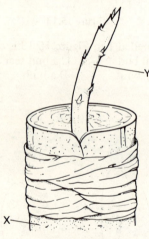

Figure 15.15

80   Figure 15.15 shows a method of propagation used for
      pear trees. What is the part labelled Y called?
      A layer   C root stock
      B stock   D scion.

81   Which of the following processes is being performed in
      Figure 15.15?
      A budding   C layering
      B grafting   D division.

82   Which one of the following stem tissues are intended to
      unite together in the method of propagation shown in
      Figure 15.15?
      A cortex   C cambium
      B epidermis   D pith.

83   Which one of the following is a feature of asexual
      vegetative reproduction?
      A one or more offspring produced
      B the parents die after reproduction
      C disease can be transmitted during reproduction
      D parents are specially adapted for the process.

84   Which one of the following is the main feature of
      vegetative and asexual reproduction?
      A gametogenesis   C mitosis
      B meiosis   D fertilization.

85   Which one of the following are prevented from forming
      in the tubers of potato when light is excluded from them?
      A starch   C chloroplasts
      B glucose   D mitochondria.

86   Which one of the following plants are unable to be
      propagated vegetatively?
      A woody perennials   C herbaceous perennials
      B biennials   D annuals.

87   Which one of the following substances are used to
      stimulate the formation of roots from cuttings?
      A starch   C sodium chloride
      B auxins   D sodium chlorate.

16

A Variety of Plants

1   A non-green or colourless plant selected from the following is:
A bracken fern   C spirogyra
B buttercup   D pin mould.

2   A plant selected from the following having a complete root, stem, and leaf system is the:
A brown seaweed   C yeast
B field mushroom   D pine tree.

3   Seeds are produced by three of the following:
(i) bracken fern, (ii) brown seaweed, (iii) toadstool,
(iv) orchid, (v) barley, (vi) common moss, (vii) dandelion,
(viii) mushroom.
A (i), (iii), (viii)   C (iv), (v), (vii)
B (ii), (iii), (vi)   D (v), (vi), (viii)

4   The smallest plant amongst the following is:
A yeast   C brown seaweed
B spirogyra   D common moss.

5   Which one of the following freshwater plants is an algae?
A duckweed   C water-lily
B pondweed   D spirogyra.

6   Phytoplankton consists mainly of:
A small fish   C algae
B shrimps   D protozoa.

7   Pin mould or mucor reproduces itself by means of:
A larvae   C seeds
B pupae   D spores.

8   A whitish-grey growth on stale bread is due to:
A protozoa   C lichens
B algae   D fungi.

9   Three examples of fungi from the following are:
(i) moss, (ii) yeasts, (iii) ferns, (iv) liverworts,
(v) mushrooms, (vi) moulds, (vii) conifers, (viii) lichens.
A (i), (iii), (iv)   C (iii), (vi), (vii)
B (ii), (v), (vi)   D (iv), (vii), (viii)

Questions 10 and 11 refer to Figure 16.1.

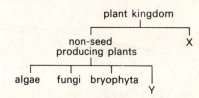

Figure 16.1

10 Which one of the following goes to X?
A mosses   C seed-bearing plants
B ferns   D seaweed.

11 Which one of the following goes to Y?
A seed-bearing plants   C moulds
B seaweed   D liverworts.

12 A rhizome is a modified:
A leaf   C root
B stem   D flower.

13 A rhizome is found in one of the following plants:
A spirogyra   C bracken fern
B buttercup   D common moss.

14 Which three of the following plants have green
coloured leaves?
(i) pin mould, (ii) toadstool, (iii) common moss,
(iv) spirogyra, (v) daffodil, (vi) pine tree.
A (i), (ii), (iv)   C (iii), (v), (vi)
B (i), (ii), (v)   D (iii), (iv), (vi)

15 Which three of the following plants have stems?
(i) meadow grass, (ii) brown seaweed, (iii) common
moss, (iv) spirogyra, (v) field mushroom, (vi) dandelion.
A (i), (iii), (vi)   C (ii), (iv), (v)
B (ii), (iii), (vi)   D (iii), (iv), (v)

16 Green coloured plants that are without roots, stems or
leaves have a vegetative body called a:
A mycelium   C thallus
B hypha   D corm.

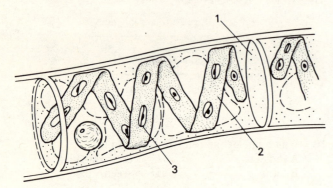

Figure 16.2

Questions **17–19** refer to Figure 16.2 which is a part of a living
spirogyra in its vegetative stage.

17 The structure labelled part 1 is a:
A hypha   C conjugation tube
B end wall   D sieve plate.

18 The structure labelled part 2 is a:
A cytoplasmic strand   C chloroplast
B male gamete   D zygospore.

19 The structure labelled part 3 is a:
A nucleus   C eye spot
B vacuole   D pyrenoid.

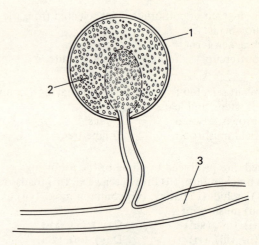

Figure 16.3

Questions **20–22** refer to Figure 16.3 which shows a stage in
the life history of a mucor.

20 What is the structure?
A sporangium   C androecium
B seed capsule   D zygospore.

21 The structures labelled part 2, inside part 1, are called the:
A spores   C pollen
B seeds   D gametes.

22 Part labelled 3 is part of the mucor plant body which
is called a:
A septate filament   C root hair
B non-septate mycelium D rhizoid.

23 Which one of the following are used as raising agents in breadmaking?
A water   C salt
B yeast   D fat.

Figure 16.4

24 Figure 16.4 is a view of the opening of a spore producing organ of one of the following:
A mucor pin mould   C common moss
B buttercup anther   D spirogyra.

25 Which one of the following is found growing and feeding on dead and decaying material?
A spirogyra   C pin mould
B rose   D seaweed.

26 Which one of the following is a coenocyte?
A fungal hyphae   C moss leaf
B spirogyra filament   D buttercup stem.

27 Lichens are found on bare rocks, and have plant bodies composed of clusters of algae surrounded by a fungus. Such a beneficial association of two different types of plants living closely together is called:
A parasitism   C commensalism
B saprophytism   D mutualism.

28 Which one of the following is an example of a saprophytic plant?
A spirogyra   C common moss
B mucor   D seaweed.

29 A green-coloured, powdery growth on wooden railings, tree trunks, and stone walls is due to:
A mosses   C algae
B moulds   D protozoa.

30 Sunlight, or artificial light, is not required for the growth of:
A fungi   C algae
B mosses   D ferns.

31 Seedlings of flowering plants if overwatered are attacked by a parasite causing 'damping off' disease. This parasite is:
A an insect   C an algae
B a fungus   D a protozoa.

32 Masses of floating spirogyra in a pond are supported by bubbles of gas. The gas is:
A air   C methane
B oxygen   D carbon dioxide.

33 When the spores of a fern plant germinate they will directly produce:
A a new fern plant
B an intermediate prothallus
C a thick coated zygospore
D a testa coated embryo.

34   Which one of the following can survive in very dry conditions for the longest period compared to the others?
A plant spores   C butterfly pupae
B house-fly larvae   D plant leaves.

35   One of the following is able to obtain its energy without using air or oxygen:
A spirogyra   C yeast
B brown seaweed   D fern.

36   'Fairy Rings', seen in meadows, and on lawns, are the result of the activity of:
A mosses   C earthworms
B fungi   D ants.

37   Which three of the following classes have members that are considered as microbes?
(i) fungi, (ii) insects, (iii) protozoa, (iv) conifers, (v) bacteria, (vi) ferns.
A (i), (iii), (v)   C (iii), (iv), (v)
B (ii), (iv), (vi)   D (iv), (v), (vi)

38   Many natural vegetable dyes, in particular *litmus*, are obtained from:
A seaweeds   C lichens
B mosses   D moulds.

39   Which one of the following help to aerate and purify sewage water?
A algae   C bacteria
B fungi   D protozoa.

40   Which of the following can prevent the growth of certain bacteria?
A peas and beans   C algae
B moulds   D fern extract.

41   The moss plant draws water and salts from the soil by means of its:
A dendrites   C rhizoids
B root hairs   D rhizome.

42   Stomata are found in two of the following plants:
(i) privet, (ii) mucor, (iii) spirogyra, (iv) mushroom, (v) seaweed, (vi) fern.
A (i), (vi)   C (iii), (v)
B (ii), (iv)   D (iv), (v)

43   The large, flattened, broad structure forming the shoot system of a fern is called a:
A compound leaf   C divided leaf
B pinnate frond   D prothallus.

44   Which two of the following conditions and locations are favoured by ferns as habitats?
(i) cool and moist, (ii) hot and dry deserts, (iii) shady woods, (iv) in pond water, (v) sandy dunes, (vi) salt mud-flats.
A (i), (iii)   C (iii), (iv)
B (ii), (iv)   D (v), (vi)

Questions 45–47 refer to Figure 16.5.

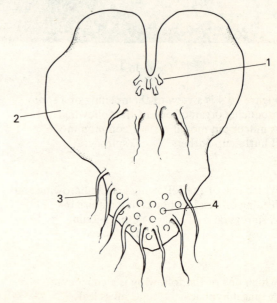

Figure 16.5

45 The green coloured, flat, heart-shaped growths shown in Figure 16.5 which are produced by the fully grown fern plant are found:
A floating in pond water
B on the upper side of the shoot system
C on the soil surface below the fern
D growing on short stalks from the shoot system.

46 The tiny growth shown in Figure 16.5 and produced by the fully formed fern plant is called a:
A seedling   C prothallus
B liverwort   D thallus.

47 Which two of the labelled structures bear the closest relationship in their function to the androecium and gynaecium of a buttercup flower?
A 1 and 2   C 1 and 4
B 2 and 3   D 4 and 3.

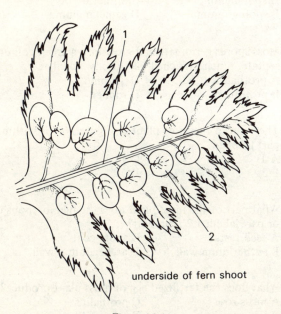

1

2

underside of fern shoot

Figure 16.6

48 What is the structure labelled part 1 in Figure 16.6 called?
A vein   C sporangium
B capillary   D rhizoid.

49 Which one of the following also possess a part similar in structure and function to that labelled 1 shown in Figure 16.6?
A mould fungus   C cabbage
B algae   D rabbit.

50 The structures labelled parts 1 and 4 in Figure 16.5 are called:
A ovary and stamen   C testis and ovaries
B archegonium and   D gametangium.
  antheridium

51 The structure labelled part 2 in Figure 16.6 is called:
A a sorus   C a sporangium
B an archegonium   D an antheridium.

52 Where are the spores produced in a fern plant?
A at 2 in Figure 16.6   C at 1 in Figure 16.6
B at 1 in Figure 16.5   D at 4 in Figure 16.5.

53 Where are the male sexual reproductive organs in a fern plant in Figure 16.5?
A 2   C 3
B 1   D 4.

54 The male sex cells of a fern are called:
A pollen   C spermatozoids
B ovules   D spermatozoa.

55 Where is the fertilized egg formed in a fern plant?
A at 2 in Figure 16.6   C at 1 in Figure 16.6
B at 1 in Figure 16.5   D at 4 in Figure 16.5.

56 What happens to the fertilized egg of a fern plant?
A it grows into sexual prothallus
B it grows into a spore forming plant
C it becomes a seed which rapidly dies
D it becomes a fleshy fruit.

57 What happens to the spores of the fern plant?
A it joins with another spore to form a seedling
B it encysts and is devoured by snails
C it germinates to become another spore-forming fern plant
D it germinates to become a prothallus.

58 Which of the following correctly summarizes the life cycle of a fern plant?
A pollen + ovule – seed – embryo – thallus – flower
B spore – thallus – spermatozoa + ovum – sporangium – seed
C prothallus – spermatozoid + egg cell – plant frond – sporangium – spore
D male and female gametangia – zygospore – sporangium – spores.

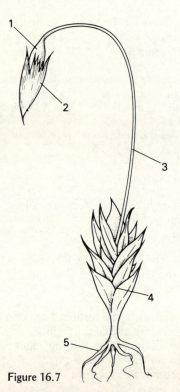

Figure 16.7

Figure 16.7 shows the external appearance of a common moss plant.

59 Where are the sexual reproductive organs located?
A 3   C 2
B 1   D 4.

60 The asexual reproductive organ in Figure 16.7 is labelled part:
A 3   C 2
B 1   D 4.

61 The asexual reproductive organ of a moss produces:
A spores   C gametes
B cormlets   D bulbils.

62 The spores of a moss are produced by the:
A sporangium   C antheridium
B sporogonium   D gametangium.

63 Moss spores germinate and grow into a green coloured septate, much branched plant body called a:
A prothallus   C rhizome
B protonema   D runner.

64 The stem of the moss plant in Figure 16.7 is within the part labelled:
A 1   C 4
B 3   D 5.

65 What part of the moss plant does the membranous cap or part labelled 2 represent?
A seed testa   C spore coat
B antheridium wall D archegonium wall.

66 What does the fertilized egg of moss plant produce?
A moss rose   C prothallus
B sporogonium   D thick-walled seed.

67   Which of the following correctly summarizes the life
     cycle of a moss plant?
     A spore – prothallus – sexual organs – fertilized egg –
        frond – sporangium
     B gametangia – gametes – zygospore – spores
     C spore – protonema – stem + leaves – male and
        female sexual organs – fertilized egg – spore capsule
     D pollen + ovules – seed – embryo – plant – flowers.

68   Which one of the following is the gametophyte generation,
     or gamete producing part, of the fern?
     A protonema   C flower
     B prothallus   D frond.

69   Which one of the following is essential in order that
     fertilization takes place in a fern?
     A insects   C wind
     B water   D warmth.

70   The spores of a moss are dispersed by means of:
     A wind   C explosive splitting of capsule
     B clinging to   D spoon action of capsule
        animal fur   teeth.

71   Each thread composing the mucor plant structure is
     called a:
     A hypha   C protonema
     B mycelium   D septate filament.

72   Which two of the following structural features are
     possessed by a filamentous green algae only, and are
     not part of a fungus?
     (i) fungal cellulose wall, (ii) chlorophyll, (iii) cellulose
     wall, (iv) glycogen, (v) spores, (vi) hyphae.
     A (ii), (iv)   C (i), (vi)
     B (ii), (iii)   D (iii), (v)

17

Microbiology and Parasites

1   Which three of the following organisms are called
microbes or micro-organisms?
(i) fleas, (ii) round-worms, (iii) bacteria, (iv) rodents,
(v) protozoa, (vi) yeasts.
A (i), (ii), (iii)   C (iii), (v), (vi)
B (ii), (iv), (v)   D (ii), (iv), (vi)

2   Which three of the following groups of organisms are
macro-organisms?
(i) bacteria, (ii) insects, (iii) viruses, (iv) fish, (v) protozoa,
(vi) round-worms.
A (i), (ii), (iii)   C (iii), (v), (vi)
B (i), (iv), (v)   D (ii), (iv), (vi)

3   Which one of the following organisms are able to feed by
photosynthesis?
A protozoa   C fungi
B viruses   D algae.

4   Which one of the following units are used to measure
microbes?
A metre   C millimetre
B centimetre   D micron.

5   A microbe which is harmful and enters the human body
to cause disease is called a:
A symbiont   C commensal
B pathogen   D saprophyte.

6   The poisonous substances produced by harmful microbes
are called:
A auxins   C toxins
B sera   D vaccines.

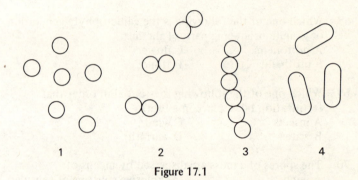

Figure 17.1

7   The bacteria labelled 1, 2, 3, and 4 in Figure 17.1 are
called:
A spirilli   C vibrios
B bacilli   D cocci.

8   The cell walls of bacteria are composed of:
A cellulose   C glycogen
B protein   D lignin.

9   Bacteria cells are without one of the following:
A nucleus   C cytoplasm
B chloroplast   D water.

10 Bacteria multiply themselves in normal conditions of
plentiful food supply by:
A seed formation   C binary fission
B spore formation   D budding.

11 Select the temperature preferred, by most bacteria causing disease in man, for normal growth and multiplication:
A 0–10 °C   C 50–60 °C
B 30–40 °C   D 90–100 °C.

12 The visible appearance of growths of bacteria seen on laboratory culture media are called:
A a thallus   C colonies
B spores   D tissues.

13 Which one of the following would be suitable as a bacteria culture media?
A dry table salt   C meat extract jelly
B dehydrated potato powder   D dry powdered milk.

14 Which two of the following foodstuffs will allow the growth of bacteria, provided suitable conditions are available?
(i) pickled onions, (ii) meat gravy, (iii) dry sultanas, (iv) dried peas, (v) fresh cream cakes, (vi) curry powder.
A (i), (iii)   C (iii), (iv)
B (ii), (v)   D (iv), (vi)

15 The non-pathogenic bacteria of man's large intestine serves to:
A digest cellulose   C produce vitamins
B produce antibodies   D produce antibiotics.

16 Which one of the following soil bacteria is a beneficial bacteria?
A denitrifying bacteria   C nitrifying bacteria
B tetanus bacteria   D typhoid bacteria.

17 Which one of the following forms of bacteria are most resistant to heat?
A cells   C spores
B colonies   D slimes.

18 Bacteria in the blood of human beings are engulfed by:
A red blood cells   C platelets
B white blood cells   D plasma.

19 One of the following are also components of the blood which serve to fight bacteria.
A allergens   C heparin
B antibodies   D haemoglobin.

20 Which one of the following fresh edible foodstuffs contains the greatest number of bacteria?
A white bread   C beer
B strawberry jam   D cheese.

21 Aerobic bacteria require one of the following gases for the purpose of respiration.
A nitrogen   C oxygen
B carbon dioxide   D ozone.

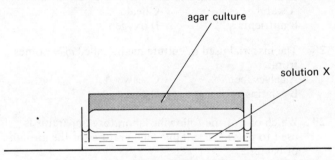

Figure 17.2

22 Figure 17.2 shows the arrangement of Petri dishes for the culture of anaerobic bacteria. What is the solution X composed of?
A boiled water   C hydrogen peroxide
B potassium pyrogallate   D lime water.

23 Which one of the following will hinder the growth of bacteria?
A vinegar   C cabbage juice
B milk   D water.

24   Which one of the following methods will sterilize a
      metal wire loop?
      A washing in tap water
      B dipping in antiseptic lotion
      C heating in a bunsen flame
      D wiping with a paper tissue.

25   The high humidity of air in a room will encourage:
      A dust to settle   C breathing to slow down
      B bacteria to multiply   D air to circulate.

26   What is the general shape of the bacteria cells
      streptococcus, staphylococcus, and diplococcus?
      A rods   C spheres
      B spirals   D cubes.

27   Which one of the following conditions is not available
      to *aerobic* bacteria in a correctly sealed can of meat?
      A water   C heat
      B nutrients   D oxygen.

28   The material used in culture media called *agar* comes
      from:
      A calves feet   C seaweed
      B gelatin   D eggs.

29   Which one of the following laboratory apparatus is
      used to provide a suitable temperature for the growth
      of bacteria?
      A refrigerator   C autoclave
      B incubator   D hot air oven.

30   Which one of the following parts of the body would be
      comparitively free from bacteria in a normal healthy
      person?
      A skin   C intestines
      B nose   D bronchioles.

31   Which one of the following parts of the body would
      contain large numbers of bacteria in a normal healthy
      person?
      A middle ear   C pleura
      B uterus   D colon.

32   Which one of the following is a bacterial disease spread
      by direct contact?
      A chicken-pox   C smallpox
      B influenza   D syphilis.

33   The purpose in providing a sample of faeces, by applicants
      for work in the catering industry, is for a doctor to
      detect whether the person:
      A suffers from diabetes
      B is a typhoid carrier
      C is anaemic
      D harbours tapeworm in the bowel.

34   Which one of the following diseases can be transmitted
      by an infected person coughing into the atmosphere?
      A lung cancer   C venereal disease
      B tuberculosis   D scabies.

35   The quarantine of cats and dogs entering Great Britain
      is for the purpose of:
      A obtaining tax
      B allow the animals to rest after long journeys
      C detecting and eliminating a specific disease
      D getting rid of their fleas.

36   Which one of the following items, which either belong
      to or are used by other people, could cause impetigo
      of the face?
      A towels   C shoes
      B books   D coins.

37   Which one of the following foods could be the cause
      of diarrhoea in a person eating the food?
      A well-heated soup   C hot coffee
      B hot potato chips   D cold ham.

38   Which one of the following is the purest and most
      bacteria-free water to drink?
      A well water   C spring water
      B refrigerator drip   D boiled tap water.
        tray water

39 Which one of the following cleaning methods keeps down airborne infection?
A broom sweeping   C feather dusters
B vacuum cleaners   D dust blowers.

40 Which one of the following officials checks that food for sale to the public is safe to eat?
A medical practitioner
B weights and measures inspector
C public health inspector
D health visitor or district nurse.

41 Bacteria in food will be killed by:
A keeping in a refrigerator
B thoroughly heating or boiling
C cooling hot food rapidly
D cooling slowly in a larder.

42 Which one of the following is a disease caused by a virus?
A pimples   C scurvy
B influenza   D tonsillitis.

43 Which one of the following is a human disease caused by a fungus?
A thread-worm   C ringworm
B tapeworm   D liver fluke.

44 A human disease caused by protozoa is:
A typhoid   C malaria
B athletes foot D scabies.

45 Which one of the following secretions of the human body can partly destroy microbes?
A saliva   C gastric juice
B bile juice   D urine.

46 Salk, Sabin, Bacille Calmette Guerin (BCG), and Jenner are names associated with one of the following:
A disinfectants C vaccines
B antiseptics   D tuberculosis.

47 Babies obtain one of the following types of immunity from their mothers:
A natural active immunity
B natural passive immunity
C artificial active immunity
D artificial passive immunity.

48 Which one of the following prevents the growth of most bacteria?
A moisture   C sunlight
B warmth   D food.

49 Which one of the following is not destroyed by boiling?
A bacteria cells   C algae cells
B bacteria spores D fern spores.

50 Antiseptics are those substances which:
A can kill all bacteria cells and spores
B can kill certain bacteria cells
C prevent the growth of bacteria
D prevent mould growth.

51 Which one of the following chemicals can be used as both an antiseptic and a disinfectant?
A glycerine   C hydrogen peroxide
B soap solution   D formaldehyde.

52 Which one of the following chemical sterilizers contain coal-tar phenols?
A iodine solution C formaldehyde
B lysol   D bleach solution.

53 Which one of the following would be the strongest disinfectant?
A eau-de-cologne   C household bleach
B pine oil fluid   D soapy water.

54 Which one of the following is the chemical ingredient of bleach solution?
A iodine   C essential oils
B phenols   D sodium hypochlorite.

55 Which of the following conditions is used to prepare 'pasteurized milk' by heating to one of the following, followed by quick cooling?
A 72 °C for 15 seconds   C 132 °C for 2 seconds
B 107 °C for 30 minutes   D 100 °C for 30 minutes.

56 Pasteurized milk is:
A sterile throughout
B free from all microbes but not spores
C free from most microbes and spores
D free from most microbes but not spores.

57 Which of the following will remove most harmful microbes from dishes and cutlery after washing in soapy water at 60 °C?
A air drying
B rinsing in water at 10 °C
C rinsing in water at 75 °C
D rinsing in water at 25 °C.

58 A baby's glass feeding bottle can be sterilized by one of the following methods:
A rinsing in water at 100 °C
B boiling in an open pan of water for 5 minutes
C heating in a pressure cooker for 15 minutes
D soaking in brine for 30 minutes.

59 The active component of sunlight which is able to destroy bacteria is:
A infra-red ray   C green light
B ultra-violet ray   D yellow light.

60 Disposable items of nursing and surgical equipment are sterilized in sealed plastic containers using one of the following:
A gamma ray   C formaldehyde
B ultra-violet ray   D infra-red ray.

61 Soiled dressings contaminated with wound discharge should be disposed of by being:
A placed in a bin   C washed and dried
B burnt in a furnace   D soaked in antiseptic.

62 Antibiotics differ mainly from all other *chemicals* which affect the growth of microbes in that they:
A destroy all microbes   C also affect the cells of the body
B destroy all viruses   D are made by certain microbes.

63 The majority of antibiotics are produced by:
A fungi   C algae
B bacteria   D artificial chemical synthesis.

64 Which one of the following conditions are essential for the growth of viruses?
A dead and decaying materials
B living cells
C presence of chlorophyll
D temperatures below freezing.

65 Which three of the following are virus diseases of man?
(i) boils, (ii) poliomyelitis, (iii) tuberculosis, (iv) measles, (v) food poisoning, (vi) mumps.
A (i), (iii), (v)   C (iii), (iv), (v)
B (ii), (iv), (vi)   D (i), (ii), (vi)

66 Viruses may be cultivated in the laboratory:
A on nutrient agar   C on humus
B in incubating eggs   D on moist bread.

67 Figure 17.3 is an ectoparasite on man called:
A a body louse   C an itchmite
B a flea   D a bed bug.

68 The ectoparasite shown in Figure 17.3 is transmitted by:
A swimming   C jumping
B flying   D crawling.

69 Which one of the following diseases is transmitted to man by the ectoparasite in Figure 17.3?
A malaria   C summer diarrhoea
B ringworm   D plague.

Figure 17.3

70 Which one of the following is the reservoir or source of the disease in Question **69**?
A kitchen refuse   C rats
B cats   D uncooked pork.

71 How is the disease introduced into man by the ectoparasite shown in Figure 17.3?
A in drinking water   C by skin bites
B on cooked food   D on unwashed fruit.

72 Which one of the following ectoparasites cements its eggs on the hair of man?
A fleas   C bed bugs
B lice   D scabies mites.

73 Which ectoparasite burrows into the skin of man to lay its eggs?
A bed bug   C body louse
B scabies mite   D pubic louse.

74 Which insect from the following is mainly responsible for food poisoning?
A butterfly   C house-fly
B bee   D ladybird.

75 Ectoparasites on man can be destroyed by spraying with:
A antiseptic   C disinfectant
B insectide   D herbicide.

76 An endoparasite of man, obtained through eating unwashed soil-contaminated vegetables, is the:
A thread-worm   C tapeworm
B ringworm   D earthworm.

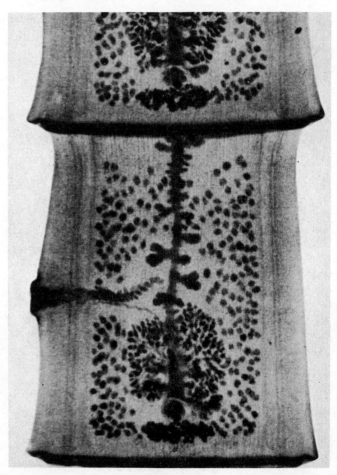

Figure 17.4

77 Figure 17.4 shows part of an endoparasite of man called a:
A round-worm   C liver fluke
B tapeworm   D bladder-worm.

78 The part of the endoparasite shown in Figure 17.4 is called the:
A proglottid   C prostomium
B peristomium   D scolex.

79 Man is infected with the endoparasite shown in Figure 17.4 through:
A cattle flea bites
B eating undercooked, infected pork
C eating trifle, contaminated by house-flies
D keeping pigeons as pets.

80 Which part of the body of man is attacked by ring-worm?
A intestine   C skin
B liver   D mouth.

81 Another name for ring-worm disease is:
A thrush   C warts
B athlete's foot   D moles.

82 The special fluid foot bath, for use in public swimming baths, is to prevent the spread of one of the following:
A foot and mouth   C scabies mite
  disease
B ring-worm   D body lice.

83 Which one of the following methods can be used to sterilize your skin?
A rinsing under a tap
B using cold cream
C swabbing with methylated spirits
D wiping with a paper tissue.

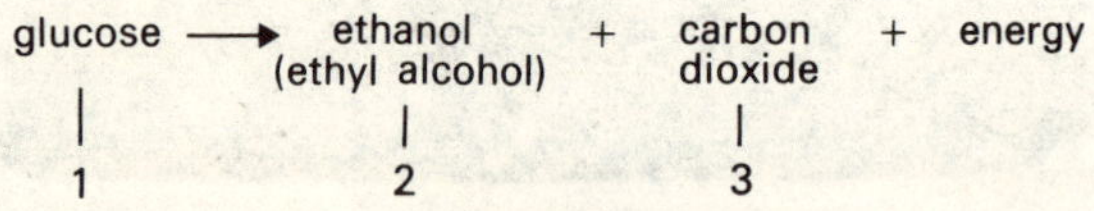

84 Which one of the following microbes can bring about the chemical change shown in Figure 17.5?
A algae   C yeast
B viruses   D bacteria.

85 Which two of the items shown in Figure 17.5 can provide a man with energy?
A 3 alone   C 1 and 2
B 1 and 3   D 2 and 3.

86 Which one of the items shown in Figure 17.5 cause the rising of flour dough when baked?
A 1   C 3
B 2   D 4.

87 Which one of the following contain the products 2 and 3 shown in Figure 17.5?
A soda water   C whisky
B beer   D brandy.

88 Yeast is a microbe belonging to one of the following:
A bacteria   C protozoa
B viruses   D fungi.

89 Which one of the following methods of infection is the cause of staphylococcal food poisoning?
A from septic skin cuts   C droppings of rats
B from the soil   D unwashed vegetables.

90 Which three of the following are food-borne diseases?
(i) measles, (ii) botulism, (iii) salmonellosis,
(iv) bronchitis, (v) typhoid, (vi) mumps.
A (i), (iii), (iv)   C (iv), (v), (vi)
B (ii), (iii), (v)   D (i), (ii), (vi)

glucose ⟶ ethanol + carbon + energy
         (ethyl alcohol)   dioxide
  | | |
  1   2   3

Figure 17.5

18

Hygiene and Health

1 Figure 18.1 shows an underground water supply. What type of water supply is shown in the figure?
A artesian well   C deep well
B shallow well   D deep spring.

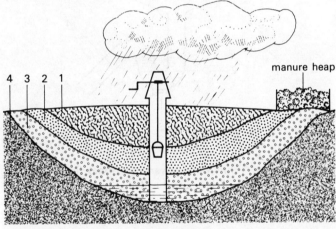

manure heap

Figure 18.1

2 The water in the well shown in Figure 18.1 will be free from any:
A microbes   C suspended matter
B soluble salts   D faecal matter.

3 Which two of the following are water-borne diseases?
(i) influenza, (ii) typhoid, (iii) malaria, (iv) lung cancer, (v) plague, (vi) cholera.
A (iii), (v)   C (i), (v)
B (ii), (i)   D (ii), (vi)

4 Which one of the following components of water are beneficial to man?
A carbon dioxide   C dissolved salts
B microbes   D dissolved air.

5 Which one of the following serve to remove harmful microbes from a mountain water reservoir?
A fish   C sand
B sunlight   D insects.

6 Rain falling on limestone hills and collecting in a reservoir, to form temporary hard water, will contain one of the following salts:
A calcium sulphate   C calcium bicarbonate
B iron sulphate   D sodium chloride.

7 Which one of the following processes *does not* take place when water is stored in a reservoir?
A sedimentation   C aeration
B filtration   D radiation.

8 Aeration is a process of water purification, which is performed mainly by one of the following:
A insects   C algae
B fish   D bacteria.

9 In the method of slow sand filtration of water, the most important layer for microbe removal is the:
A large gravel   C sand
B green slime   D perforated brick.

10 The final process of large-scale water purification
following filtration is called:
A fluoridation   C distillation
B chlorination   D oxidation.

11 Which of the following is *least* suited in its effect on the
environment in providing a water supply for a chemical
industry?
A construct a reservoir in a nature reserve
B desalination of sea water
C distillation of river water by solar energy
D utilization of purified sewage water.

12 Which one of the following is the greatest cause of
pollution in a country lake?
A swimming and boating by tourists
B campers and caravaners on lake shores
C fertilizers washed off surrounding farm land
D contamination by seagulls and waterfowl.

13 Small quantities of water can be rapidly purified for
drinking purposes by:
A adding sodium chloride tablets
B adding sodium hypochlorite tablets
C leaving in the sun for 10 minutes
D blowing through a straw by mouth.

14 The fluid used in a chemical lavatory of a caravan or
aircraft is:
A soapy water   C formaldehyde
B sulphuric acid   D brine.

15 Hospital dry refuse, in the form of soiled dressings etc.,
is disposed of by:
A flushing down the hospital drains
B local authority refuse collectors
C burning on the hospital premises
D tipping, spraying and burial in hospital grounds.

16 Waste from ships is thrown into the sea. Which one of
the following rubbish items will not undergo
decomposition in sea-water?
A plastic containers   C paper bags
B tin cans   D vegetable peelings.

17 Household refuse should be disposed of by one of the
following methods, in order to prevent the breeding of
house-flies.
A emptied into metal dustbins with loose-fitting lids
B wrapped in newspaper and put in metal bins with
    loose-fitting lids
C wrapped in newspaper and put in paper sacks
D emptied into open top paper sacks.

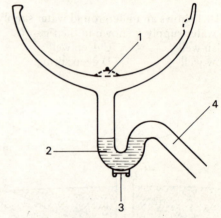

Figure 18.2

18 Figure 18.2 shows a transverse sectional view of a sink.
What is the part labelled 2 called?
A an overflow   C a strainer
B a water seal   D a siphon.

19 The purpose of part labelled 2 in Figure 18.2 is to:
A empty the sink by suction
B prevent entry of drain gases
C collect insoluble waste
D allow fresh air to enter.

20 Which one of the following methods can be used to clear
a blockage from a sink of the type shown in Figure 18.2?
A insert a wire down 1
B insert a rod up 4
C unscrew 3 and insert a wire
D unscrew 1 and disconnect 2.

21 Which one of the following *do not* empty their waste into a grated gully trap?
A kitchen sink   C bath
B water-closet pan   D rain-water pipe.

22 Which one of the following is without a S or V bend in its emptying pipe?
A bath and wash basin   C roof gutter
B bidet   D kitchen sink.

23 Which of the following is the correct order in which waste water leaves a wash hand basin to reach the sewage works?
A sewer – wastepipe – inspection chamber – gulleytrap – house drain
B waste pipe – gulley traps – inspection chamber – house drain – sewer
C house drain – inspection chamber – waste pipe – sewer – gulley traps
D gulley trap – house drain – waste pipe – inspection chamber – sewer.

24 Which of the following is the correct order in which sewage is treated at the main sewage purification works?
A settlement – screening – sedimentation – aeration
B aeration – sedimentation – settlement – screening
C screening – sedimentation – aeration – settlement
D sedimentation – screening – aeration – settlement.

25 The process of screening removes:
A faeces and urine   C microbes
B rags and sticks   D sand and gravel.

26 Which three of the following are most active in purifying the effluent in the aeration stage?
(i) algae, (ii) anaerobic bacteria, (iii) fungi, (iv) nitrogen, (v) aerobic bacteria, (vi) oxygen.
A (i), (ii), (iii)   C (iv), (v), (vi)
B (ii), (iii), (iv)   D (i), (v), (vi)

27 The dark brown muddy liquid in the sewage settlement tanks produces a clean effluent water and one of the following:
A peat   C sludge
B humus   D grit.

28 If a main house drain is found to be blocked, which of the following will attend to it for you?
A The Water Board   C The Police
B Public Health   D The Gas Board.
  Authority

29 Which one of the following forms the best site for building a house?
A low-lying clay   C tipped soil
B elevated granite   D sandy soil.

30 A cool larder in a house should be located facing one of the following points:
A North   C East
B South   D West.

31 In which direction should a house face to receive the most sun?
A North   C East
B South   D West.

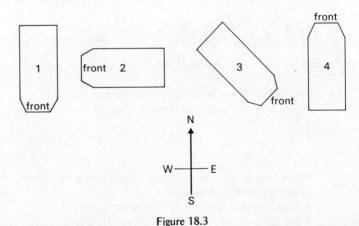

Figure 18.3

32   Which diagram in Figure 18.3 is the best aspect for siting a house in order that all four walls benefit from the sun's rays?
A 1   C 3
B 2   D 4.

33   Rising damp in walls is due to:
A a defective roof   C crack in side wall
B no damp-proof course D broken gutters.

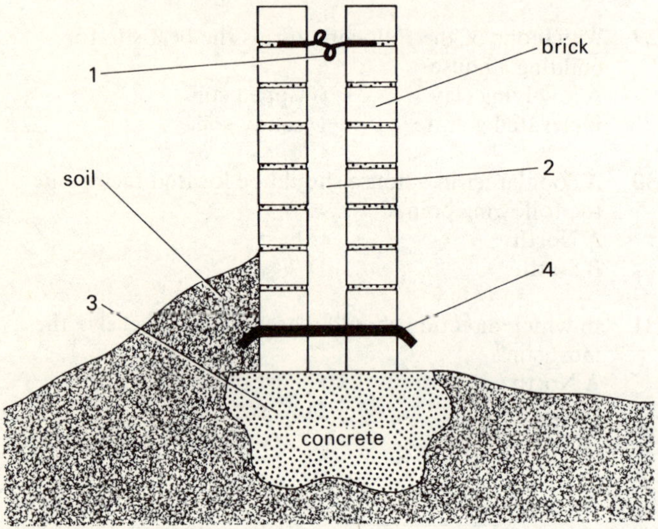

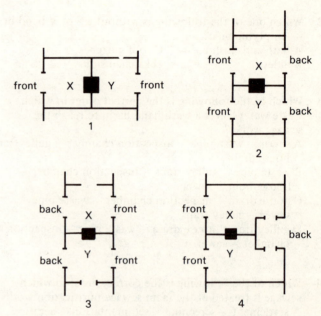

Figure 18.4

34   Which part of the brick wall construction shown in Figure 18.4 will act as an insulator of heat?
A 1   C 3
B 2   D 4.

35   Which one of the following will flourish in dark, damp, airless spaces below a floor?
A wood-worm   C death-watch beetle
B dry rot   D furniture beetle.

36   Which one of the following types of lighting will destroy microbes?
A candlelight   C gas light
B electric light   D sunlight.

Figure 18.5

37   Figure 18.5 shows the plans of two adjoining houses X and Y. Which of the house plans is considered unfit for human habitation?
A 1   C 3
B 2   D 4.

38   The reason why the house plan in Figure 18.5 is unsuitable for human habitation is:
A one house will receive more sunlight than the other
B one house will be exposed to prevailing wind and rain more than the other
C the houses cannot be properly ventilated
D houses built in a row would be without gardens.

39 Which of the following instruments is used to measure the humidity of the air?
A thermometer
B hygrometer
C barometer
D hydrometer.

40 The comfortable temperature of a sitting-room should be:
A 8 °C
B 18 °C
C 28 °C
D 82 °C.

41 Which of the following methods of heating does not alter the chemical composition of room air?
A electric fires
B gas fires
C oil heaters
D coal fires.

42 The normal percentage composition of air is:

	Nitrogen	Oxygen	Carbon dioxide	Rare gases
A	0.03	20	78	1
B	78	20	0.03	1
C	1	0.03	20	78
D	20	78	1	0.03

Use the following to answer Questions 43 and 44.
(i) sand, (ii) carbon monoxide, (iii) microbes, (iv) lead, (v) sulphur dioxide, (vi) pollen, (vii) soot, (viii) asbestos dust.

43 Which three of the substances in the list above are air pollutants produced from motor cars and traffic in cities?
A (i), (iii), (vi)
B (ii), (iv), (viii)
C (ii), (iv), (v)
D (v), (vii), (viii)

44 Which two substances from the list above are air pollutants produced by burning coke, oil, coal and gas fuels in a good supply of air, for domestic heating and industrial power?
A (i), (ii)
B (iii), (viii)
C (iv), (vi)
D (v), (vii)

45 Where does asbestos dust air pollution come from?
A car tyres
B car brakes
C household chimneys
D aircraft.

46 Which one of the following diseases is increased by air pollution?
A rheumatism
B deafness
C influenza
D bronchitis.

47 The production of chimney smoke in the United Kingdom is prohibited:
A everywhere in town and countryside
B in London only
C in all town and cities
D in some towns and cities.

48 The amount of carbon dioxide in the air in recent years has:
A remained the same
B increased by 10 per cent
C decreased by 10 per cent
D decreased by 20 per cent.

49 Changes in the composition of the air by supersonic aircraft may be due to:
A supersonic boom
B oil pollution
C oxygen consumption
D increasing microbe content.

50 Which one of the following is the most perishable food?
A raspberry jam
B dehydrated potato
C smoked salmon
D milk.

51 The least perishable food amongst the following are:
A fish
B eggs
C sultanas
D lettuces.

52 One of the following foods cannot be preserved by pasteurization:
A milk
B eggs
C wine
D orange juice.

53   Pasteurization is the process of heating a food for a short time at one of the following temperatures:
A 120 °C   C 100 °C
B 70 °C   D 95 °C.

54   Which one of the following is liable to Salmonella food poisoning infection?
A cheese   C duck eggs
B biscuits   D pickled onions.

55   The process of boiling foods in open pans is unable to destroy the spores of one of the following microbes:
A salmonella   C staphylococci
B Clostridium welchii   D streptococcus.

56   Which one of the following processes will destroy most bacteria in meat?
A mincing   C hanging
B roasting   D smoking.

57   When carrots are preserved by canning in plain water they are protected from bacterial spoilage by two of the following:
(i) chemical preservative, (ii) acids, (iii) vacuum, (iv) high temperature, (v) freeze drying, (vi) antibiotics.
A (i), (ii)   C (iii), (iv)
B (ii), (v)   D (iv), (vi)

58   A packet of processed dried peas contains a chemical preservative called:
A sucrose   C sulphur dioxide
B brine   D acetic acid.

59   Milk powder, spaghetti, tea, and coffee extract are all preserved by:
A smoking   C chemical preservation
B dehydration   D atomic radiation.

60   Which one of the following is a preservative in cheese?
A sugar   C carbon dioxide
B microbes   D casein.

61   Which one of the following ingredients of canned and bottled fruits is the main preservative?
A sugar   C artificial colouring
B water   D citric acid.

62   Which one of the following ingredients of a bottle of pickled onions is the main preservative?
A caramel   C acetic acid
B spices   D sodium chloride.

63   Which one of the following foods would *not* contain a preservative?
A soft drinks   C packed lard
B marzipan   D fresh milk.

64   Which one of the following is a useful preservative for sardines?
A tomato juice   C sugar syrup
B olive oil   D benzoic acid.

65   Saltpetre, or potassium nitrate, is a preservative for:
A beer   C pork
B cereals   D apricots.

66   Salt, or sodium chloride, preserves food by withdrawing water from it by:
A diffusion   C evaporation
B osmosis   D capillarity.

67   When water is removed from a frozen food by sublimation under a vacuum, the process is called:
A vacuum drying   C freeze drying
B spray drying   D deep freezing.

68   If beef has been preserved by the process in Question 67, it can be prepared for roasting:
A by thawing   C without treatment
B by adding water   D by sprinkling with flour.

69   The temperature in a deep-freezing cabinet should be:
A −20 °C   C 0 °C
B 20 °C   D −200 °C.

70   Which one of the following foods should *never* be
      stored in a refrigerator?
      A fish   C bananas
      B cheese   D eggs.

71   Which one of the following methods of food
      preservation, now under investigation, would have the
      least harmful effect on man?
      A antibiotics   C ultra-violet radiation
      B atomic radiation   D chemical disinfection.

72   Which one of the following types of treated milk will
      keep the longest, at room temperature in unopened
      containers?
      A sterilized milk   C pasteurized milk
      B homogenized milk   D ultra-heat treated milk.

73   Which one of the following is the preservative added
      to condensed milk?
      A sulphur dioxide   C benzoic acid
      B sugar   D calcium phosphate.

74   Which one of the following diseases of man can be due
      to drinking untreated milk from an infected cow?
      A diphtheria   C anthrax
      B tuberculosis   D typhoid.

75   People who handle or process milk in dairies should not
      work if they are suffering from one of the following:
      A toothache   C sore throat
      B ear-ache   D warts on hands.

76   Dairy workers who smoke can cause infection of milk
      through:
      A the cigarette ash   C their fingers
      B tobacco smoke   D dead matches.

77   All workers who handle food should make certain that
      all cuts and abrasions on their hands are:
      A washed and left uncovered
      B covered with a cotton bandage
      C covered by a cotton flesh coloured adhesive dressing
      D covered by a brightly coloured waterproof adhesive
         dressing.

78   After visiting the toilet and washing their hands, food
      workers should then use the following to dry their hands.
      A a short roller towel   C a continuous roller towel
      B a disposable paper   D a square cotton hand towel.
         towel

79   Foodworkers wear a hat or cap to:
      A look attractive in uniform
      B protect their hair from condensation from ceilings
      C completely cover the hair without any being visible
      D prevent head and hair infections.

80   Catering supervisors should particularly check that all
      staff handling food have *one* of the following:
      A attractive hair styles
      B short clean finger nails
      C a well-pressed uniform
      D clean shoes.

81   Which one of the following officials check kitchens of
      restaurants and cafes for cleanliness?
      A Health Visitors
      B Weights and Measures Inspectors
      C Public Health Inspectors
      D Family Doctors.

82   Which one of the following parts of the human body are
      places *where* most bacteria will breed?
      A nostrils   C hair
      B eyes   D skin surface.

83   Which one of the following is a disease due to the
      pollution of air?
      A stomach ulcer   C bronchitis
      B alcoholism   D rheumatism.

84   The disease which shortens the life of fat, inactive
      people is:
      A appendicitis   C tuberculosis
      B heart disease   D malnutrition.

85   Cigarette smoking is without doubt one of the causes of:
      A tuberculosis   C lung cancer
      B diabetes   D venereal disease.

86 The one main cause of obesity is:
    A lack of exercise   C gland disorders
    B overeating   D drinking too much water.

87 The main cause of death in poor countries is:
    A obesity   C heart disease
    B malnutrition   D cancer.

88 Which three of the following lead to mental disorders
    and insanity?
    (i) overeating, (ii) venereal disease, (iii) drug addiction,
    (iv) alcoholism, (v) lack of exercise, (vi) cigarette
    smoking.
    A (i), (ii), (iii)   C (ii), (iv), (vi)
    B (ii), (iii), (iv)   D (i), (v), (vi)

89 Which one of the following have led to an increase in
    venereal disease?
    A religious reasons   C alcoholic indulgence
    B fear of pregnancy   D moral reasons.

90 Which one of the following diseases has been conquered
    by modern medical science in the United Kingdom?
    A cancer   C heart disease
    B tuberculosis   D rheumatism.

Answers

CHAPTER 1 (page 1)

1 A	2 C	3 B	4 A	5 B	6 C	7 A	8 B
9 A	10 A	11 B	12 B	13 C	14 B	15 A	16 C
17 B	18 D	19 A	20 D	21 C	22 A	23 A	24 C
25 A	26 C	27 B	28 A	29 A	30 C	31 A	32 B
33 A	34 C	35 D	36 A	37 C	38 C	39 B	40 D
41 C	42 C	43 B	44 A	45 D	46 A	47 D	48 B
49 A	50 C.						

CHAPTER 2 (page 5)

1 B	2 A	3 A	4 C	5 B	6 A	7 B	8 D
9 B	10 C	11 A	12 C	13 A	14 D	15 C	16 C
17 B	18 A	19 D	20 A	21 B	22 C	23 C	24 C
25 A	26 D	27 D	28 D	29 A	30 A	31 B	32 D
33 B	34 A	35 D	36 A	37 C	38 B	39 D	40 B
41 A	42 C	43 A	44 B	45 C.			

CHAPTER 3 (page 10)

1 D	2 A	3 B	4 A	5 C	6 D	7 B	8 C
9 D	10 B	11 C	12 C	13 A	14 D	15 B	16 D
17 B	18 D	19 B	20 C	21 B	22 A	23 B	24 B
25 A	26 B	27 D	28 D	29 B	30 C	31 A	32 A
33 C	34 A	35 A	36 D	37 B	38 A	39 B	40 A
41 B	42 B	43 D	44 A	45 C	46 A	47 C	48 C
49 B	50 C	51 A	52 B	53 C	54 C	55 B	56 C
57 C	58 B	59 A	60 B	61 C	62 A	63 B	64 D
65 D	66 B	67 C	68 B	69 D	70 D	71 B	72 A
73 C	74 D	75 B	76 C.				

CHAPTER 4 (page 16)

1 B	2 B	3 A	4 B	5 C	6 D	7 A	8 C
9 D	10 A	11 A	12 C	13 C	14 D	15 A	16 B
17 D	18 C	19 C	20 C	21 C	22 B	23 D	24 C
25 B	26 C	27 B	28 A	29 C	30 C	31 B	32 C
33 C	34 A	35 B	36 C	37 C	38 A	39 C	40 B
41 C	42 A	43 C	44 D	45 A	46 D	47 C	48 D
49 D	50 C	51 B	52 B	53 D	54 B	55 B	56 C
57 B	58 B	59 C	60 A	61 A	62 C	63 B	64 C
65 B	66 A	67 D	68 A	69 A	70 D	71 D	72 B
73 A	74 C.						

CHAPTER 5

Respiration (page 23)

1 C	2 D	3 B	4 A	5 A	6 C	7 A	8 D
9 B	10 D	11 B	12 A	13 B	14 A	15 B	16 C
17 C	18 C	19 C	20 B	21 C	22 C	23 B	24 C
25 A	26 B	27 C	28 D	29 C	30 B	31 B	32 D
33 B	34 A	35 B	36 C	37 C	38 C	39 B	40 B
41 C	42 A	43 B.					

Excretion (page 27)

1 C	2 B	3 B	4 C	5 C	6 C	7 C	8 B
9 C	10 B	11 C	12 A	13 A	14 C	15 C	16 A
17 B	18 C	19 A	20 C	21 D	22 A	23 C	24 B
25 B	26 C	27 B	28 C	29 D	30 C.		

Skin (page 30)

1 D	2 B	3 A	4 C	5 B	6 B	7 B	8 C
9 D	10 B	11 C	12 D	13 B	14 D	15 C	16 B
17 A	18 B	19 C	20 C	21 D	22 A	23 D	24 B
25 A	26 D	27 A	28 A	29 B	30 B	31 C.	

CHAPTER 6 (page 33)

1 C	2 A	3 B	4 C	5 C	6 A	7 B	8 A
9 B	10 C	11 A	12 B	13 B	14 C	15 B	16 C
17 A	18 D	19 C	20 A	21 C	22 B	23 A	24 A
25 B	26 A	27 C	28 B	29 B	30 B	31 C	32 A
33 B	34 C	35 C	36 D	37 A	38 A	39 D	40 C
41 C	42 B	43 B	44 A	45 D	46 A	47 D	48 D
49 B	50 D	51 C	52 B	53 B	54 A	55 B	56 A
57 A	58 D	59 D	60 B	61 B	62 C	63 C	64 B
65 A	66 A	67 C	68 A	69 B	70 D	71 B	72 B
73 C	74 C	75 B	76 B	77 C	78 C	79 A	80 A
81 B	82 B	83 A.					

CHAPTER 7 (page 41)

1 D	2 C	3 B	4 C	5 B	6 D	7 A	8 B
9 B	10 A	11 A	12 A	13 A	14 B	15 C	16 A
17 B	18 B	19 B	20 B	21 D	22 B	23 D	24 C
25 B	26 D	27 B	28 B	29 B	30 B	31 C	32 B
33 C	34 C	35 B	36 D	37 A	38 A	39 C	40 C
41 C	42 D	43 C	44 C	45 C	46 A	47 C	48 B
49 D	50 C	51 B	52 C	53 B	54 A	55 B	56 B
57 D	58 B	59 C	60 B	61 B	62 C	63 C	64 C
65 C	66 C	67 A	68 D	69 D	70 C	71 B	72 A
73 D	74 C	75 C.					

CHAPTER 8 (page 49)

1 A	2 C	3 A	4 B	5 B	6 C	7 B	8 C
9 D	10 D	11 D	12 C	13 B	14 D	15 D	16 A
17 C	18 C	19 C	20 C	21 B	22 B	23 A	24 B
25 A	26 C	27 B	28 D	29 B	30 D	31 A	32 B
33 B	34 D	35 C	36 A	37 C	38 B	39 B	40 B
41 C	42 B	43 B	44 A	45 C	46 C	47 C	48 C
49 B	50 C	51 B	52 B	53 C	54 B	55 D	56 B
57 D	58 C	59 D	60 B	61 B	62 A	63 B	64 D.

CHAPTER 9 (page 56)

1 B	2 D	3 A	4 B	5 C	6 A	7 B	8 D
9 A	10 B	11 B	12 B	13 B	14 B	15 D	16 D
17 D	18 D	19 C	20 C	21 B	22 C	23 B	24 B
25 C	26 C	27 C	28 A	29 C	30 C	31 B	32 B
33 D	34 B	35 B	36 C	37 B	38 D	39 A	40 C
41 B	42 B	43 C	44 A	45 B	46 A	47 B	48 B
49 D	50 D	51 B	52 B	53 B	54 B	55 A	56 D
57 C	58 C	59 A	60 C	61 D	62 B	63 C	64 C
65 C	66 C	67 B	68 D	69 B	70 D	71 A	72 C
73 C	74 C.						

CHAPTER 10 (page 62)

1 D	2 B	3 D	4 C	5 C	6 C	7 B	8 B
9 B	10 C	11 C	12 C	13 C	14 C	15 A	16 B
17 B	18 B	19 D	20 C	21 C	22 D	23 C	24 A
25 B	26 B	27 A	28 D	29 A	30 A	31 D	32 A
33 B	34 D	35 B	36 D	37 A	38 C	39 C	40 D
41 C	42 B	43 A	44 B	45 B	46 D	47 B	48 B
49 C	50 C	51 B	52 B.				

CHAPTER 11 (page 68)

1 B	2 C	3 D	4 D	5 C	6 A	7 D	8 B
9 D	10 A	11 C	12 B	13 A	14 C	15 D	16 C
17 D	18 B	19 B	20 B	21 A	22 A	23 D	24 B
25 B	26 B	27 B	28 B	29 B	30 C	31 B	32 B
33 C	34 B	35 C	36 C	37 B	38 B	39 B	40 C
41 C	42 C	43 D	44 B	45 A	46 C	47 B	48 A
49 C	50 C	51 A	52 B	53 B	54 C	55 B	56 C
57 A	58 B	59 C	60 D	61 B	62 B.		

CHAPTER 12 (page 74)

1 C	2 B	3 A	4 D	5 B	6 C	7 C	8 B
9 A	10 B	11 C	12 D	13 B	14 C	15 C	16 D
17 C	18 D	19 A	20 C	21 B	22 C	23 A	24 C
25 B	26 D	27 D	28 B	29 A	30 B	31 B	32 A
33 B	34 C	35 C	36 B	37 B	38 C	39 C	40 C
41 C	42 B	43 B	44 C	45 B	46 B	47 B	48 B
49 B	50 B	51 C	52 A	53 D	54 C	55 A	56 B
57 B	58 B	59 A	60 B	61 B	62 D	63 C	64 B
65 C	66 C	67 B	68 D	69 B	70 A	71 C	72 A
73 D	74 B	75 D	76 B	77 C	78 A	79 A	80 B
81 B	82 B	83 C	84 B	85 A	86 C	87 B	88 A
89 C	90 A	91 B	92 C	93 C	94 C	95 C	96 B
97 C	98 C	99 A	100 B	101 D	102 A	103 A	104 D
105 A	106 C.						